Lecture Notes in Computer Science 1343

Lecture Notes in Computer Science   1343
Edited by G. Goos, J. Hartmanis and J. van Leeuwen

Advisory Board: W. Brauer   D. Gries   J. Stoer

Springer
Berlin
Heidelberg
New York
Barcelona
Budapest
Hong Kong
London
Milan
Paris
Santa Clara
Singapore
Tokyo

Yutaka Ishikawa   Rodney R. Oldehoeft
John V.W. Reynders   Marydell Tholburn (Eds.)

Scientific Computing in Object-Oriented Parallel Environments

First International Conference, ISCOPE 97
Marina del Rey, California, USA
December 8-11, 1997
Proceedings

 Springer

Volume Editors

Yutaka Ishikawa
Real World Computing Partnership
Tsukuba Mitsui Building 16F
1-6-1 Takezono, Tsukuba-shi, Ibaraki 305, Japan
E-mail: ishikawa@trc.rwcp.or.jp

Rodney R. Oldehoeft
Computer Science Department, Colorado State University
Fort Collins, CO 80523, USA
E-mail: rro@cs.colostate.edu

John V.W. Reynders
Advanced Computing Lab, MS B287, Los Alamos National Laboratory
Los Alamos, NM 87545, USA
E-mail: reynders@acl.lanl.gov

Marydell Tholburn
CIC-8, MS B272, Los Alamos National Laboratory
Los Alamos, NM 87545, USA
E-mail: marydell@lanl.gov

Cataloging-in-Publication data applied for

Die Deutsche Bibliothek - CIP-Einheitsaufnahme

Scientific computing in object oriented parallel environments :
first international conference ; proceedings / ISCOPE 97, Marina del
Rey, California, December 8 - 11, 1997. Yutaka Ishikawa ... (ed.). -
Berlin ; Heidelberg ; New York ; Barcelona ; Budapest ; Hong Kong
; London ; Milan ; Paris ; Santa Clara ; Singapore ; Tokyo : Springer,
1997
   (Lecture notes in computer science ; Vol. 1343)
   ISBN 3-540-63827-X

CR Subject Classification (1991): D.1, D.3, D.2, G.1-2

ISSN 0302-9743
ISBN 3-540-63827-X Springer-Verlag Berlin Heidelberg New York

© Springer-Verlag Berlin Heidelberg 1997
Printed in Germany

Typesetting: Camera-ready by author
SPIN 10652621   06/3142 – 5 4 3 2 1 0   Printed on acid-free paper

Introduction

Object-oriented software design has had a profound impact on the computing industry. The basic concepts of object encapsulation, type abstraction, polymorphism and inheritance have become the foundation of the desktop software revolution, and they are at the core of the coming generation of Internet web technologies. The reasons for this are simple to understand. Today's software must interoperate transparently, and must be "network aware." In addition, these applications must provide greater end-user programmability and allow extensive, discipline-specific specialization. Driven by the need to build these increasingly complex programs in shorter development cycles, developers have had to turn away from monolithic designs and learn to build new systems by composing and extending existing standard components. An object-oriented software industry called "middleware" has emerged to make this possible. Object technology such as ActiveX, CORBA, C++, and Java forms the foundation of this new programming infrastructure.

At first glance, it may seem that the world of scientific and engineering application design has missed this basic paradigm shift that the rest of the software industry has embraced. Supercomputer vendors have gone from vector machines to massively parallel, distributed-memory systems, and now to large-scale, shared-memory multiprocessors, but the software technology remains mired in the 20-year-old Fortran codes that ran on the Cray 1. However, a closer inspection will reveal a different story. Scientific programmers are driven by the same pressures to produce interoperable, scalable, user-programmable, "network aware" applications. Further, because the lifetime of many supercomputer architectures is far shorter than the lifetime of many large-scale applications, software designers have had to find ways in which the level of abstraction and expression of scientific computation can be made independent of a particular computing platform, while still taking advantage of specific architectural features for high performance. While there are several ways to accomplish this, object-oriented approaches seemed natural and, in the late 1980s, scientific software researchers began experimenting with this approach. In 1993, a pioneering group at Rogue Wave Software decided to hold a small conference on object-oriented numerics at a ski resort. The organizers expected a few submissions and only 20 participants. Instead, they attracted dozens of papers and well over 100 attendees at their meeting, which they called OON-SKI. Few people came for the skiing, so the name was changed to OONSCI to reflect the focus on science. The Fourth OONSCI Conference was held at Mississippi State University in March, 1996.

About the same time that OON-SKI was being created, work on object-oriented methods for scientific applications was starting at the U.S. national laboratories. In 1994, a group at Los Alamos decided it was time to hold a meeting that would focus on "Parallel Object-Oriented Methods and Applications," and in December, 1994, the first POOMA conference was held in Santa Fe, New Mexico. As with OONSCI, the response to the POOMA meeting was excellent. The second POOMA meeting was held in Santa Fe in February, 1996.

In Japan, the Real World Computing Partnership (RWCP) had by this time begun a large and exciting project that used a variety of object-oriented software technologies as the foundation for a large system design effort. One of their focus projects was on parallel extensions to C++ and the associated compiler technology. After a series of small meetings with people from Los Alamos and the HPC++ group in the U.S., and the European ESPRIT Europa working group, RWCP hosted the first "International Workshop on Parallel C++" (IWPC++) in March, 1996 in Kanazawa, Japan. It focused exclusively on C++ issues in parallel computation.

In November, 1996, the organizers of OONSCI, POOMA, and IWPC++ met and decided to establish an international forum that would combine the objectives of all three conferences. The result is the International Scientific Computing in Object-Oriented Parallel Environments Conference (ISCOPE). This volume contains edited versions of the papers selected for presentation at ISCOPE'97, which was held in Marina del Rey, California in December, 1997.

The papers can be organized into six categories. The first set treats the subject of run-time performance optimization at several levels. The first paper, by Stephen J. Fink and Scott B. Baden, addresses the design of adaptive, block-structured grids on clusters of SMP servers. The second paper, by Nobuhisa Fujinami, considers the lower levels of run-time optimization by considering run-time code generation for C++. Federico Bassetti, Dan Quinlan, and Kei Davis consider several approaches to the important problem of optimizing array classes. In the last paper in this group, Matthias Weidmann considers the impact on performance due to the use of the C++ standard template library on an algebraic multi-grid code.

The second group of eleven papers considers new language and programming paradigms. Andrew Chien, Julian Dolby, Bishwaroop Ganguly, Vijay Karamcheti, and Xingbin Zhang describe how Illinois Concert C++ (ICC++) can be used for irregular applications. Jaakko Järvi treats the problem of sparse vectors at compile time. Todd Veldhuizen and M. Ed Jernigan provide an excellent analysis of code generation problems and optimizations that are possible in C++ in their paper "Will C++ Be Faster than Fortran?". MPI is a critical library for parallel programming and MPI-2 is now well-defined. Jeff Squyres, Bill Saphir, and Andrew Lumsdaine describe the issues related to building a C++ interface to this important standard.

Of course C++ is not the only language for parallel, object-oriented programming. Jürgen Quittek and Boris Weissman describe the mechanisms for synchronization in the Sather language. Wouter Joosen, Bert Robben, Henk Van Wulpen, and Pierre Verbaeten describe CORRELATE and illustrate its use with a molecular dynamics simulation.

Satoshi Matsuoka, N. Nikami, H. Ogawa, and Y. Ishikawa describe the use of MTTL and MPC++ in the implementation of a variety of parallel programming paradigms without using extensions to the base language. Tosiyuki Takahashi, Yutaka Ishikawa, Mitsuhisa Sato, and Akinori Yonezawa provide an excellent description of the MPC++ metalevel compiler architecture. Matthias Besch,

Hua Bi, Gerd Heber, Matthias Kessler, and Matthias Wilhelmi describe Promoter, a topology-based coordination language for data parallelism. In another RWCP collaboration, Jens Gerlach, Mitsuhisa Sato, and Yutaka Ishikawa describe how to use STL with a templated library of iterative methods and the Promoter run-time system to build adaptive finite element programs. Motohiko Matsuda, Mitsuhisa Sato, and Yutaka Ishikawa describe the use of STL adaptors to implement an array class.

Java is having no less an impact on scientific computation than in the rest of the industry. Five papers describe the application of Java-based technology on problems of interest to this community. Gerald Löffler describes the use of multi-threading for the solution of PDEs. Sava Mintchev and Vladimir Getov discuss the problem of binding scientific libraries to native methods in Java. Kenji Imasaki describes how PVM can be integrated with the Java environment. Ju-Pin Ang and Yong-Tai Tan describe how Java RMI and other technology can be used to implement a mobile computing framework. Finally, Yuuji Ichisugi and Yves Roudier describe a tool for extending and preprocessing Java programs. They show how it can be used to build a data parallel extension to the language.

Five papers deal directly with applications. Are Magnus Bruaset, Xing Cai, Hans Petter Langtangen, and Aslak Tveito describe the concept of sequential simulators for parallel solution of PDEs. C. Calvin and Ph. Emonot describe a parallel, object-oriented thermal-hydraulic 3-D code. David Brown, Bill Henshaw, and Daniel Quinlan write about the features of the Overture class library for solving PDEs. William Humphrey, Steve Karmesin, Federico Bassetti, and John Reynders discuss the POOMA framework, which is an important part of the ASCI initiative. In a related ASCI project, Julian Cummings, Stephen Lee, Steven Nolen, and Noel Keen describe an object-oriented application, MC++, to find the eigenvalues of the neutron transport equation.

A set of six papers discuss the role of object-oriented libraries. Brian McCandless and Andrew Lumsdaine describe the STL-inspired matrix template library and the role of poly-algorithms. The design of data field classes for multi-physics applications is described by Takashi Ohta. Wei-Min Jeng and Camilliam Lin discuss parallel image reconstruction techniques for positron emission tomography, and M.E. Henderson and S.E. Lyons describe reservoir simulation. Donald Bashford presents the design of a class library for molecular electrostatic models for analyzing biological molecules. Finally, Cecelia DeLuca, Curtis W. Heisey, Robert A. Bond, and Jim M. Daly describe a class library for real-time radar signal processing based on space-time adaptive processing algorithms.

The final set of papers describes a collection of new ideas and approaches to parallel scientific computation. John Irwin, Jean-Marc Loingtier, John Gilbert, Gregor Kiczales, John Lamping, Anurag Mendhekar, and Tatiana Shpeisman describe a model called "aspect-oriented" programming, in which the algorithmic components of a computation are separated from the "aspects" associated with putting the components together in an efficient solution. Russell F. Haddleton and John L. Pfaltz describe the design of a "data parallel" object-oriented data base. Pattern-based approaches to the design of parallel computations is the subject of a paper by Steve MacDonald, Jonathan Schaeffer, and Duane Szafron. Paul Gray and Vaidy Sunderam describe the IceT system, a Java-based, distributed, collaborative environment for parallel programming. Ravi Ramamoorthi, Adam Rifkin, Boris Dimitrov, and K. Mani Chandy conclude this volume with an excellent paper on the general problem of resource reservation in heterogeneous distributed environments.

This collection of papers represents the state of the art in the application of object-oriented methods in scientific and engineering applications. The conference and its subject area are truly international in scope, with authors participating from 10 countries. The ISCOPE organizers are confident that the reader will share our excitement about this dynamic area of computer science and application research.

Organizing Committee

Scott Baden, University of California at San Diego, USA
Denis Caromel, INRIA Sophia Antipolis, France
Dennis Gannon, University of Indiana, USA
Yutaka Ishikawa, Real World Computing Partnership, Japan
Carl Kesselman, University of Southern California, USA
Satoshi Matsuoka, Tokyo Institute of Technology, Japan
Rodney R. Oldehoeft, Colorado State University, USA
John V. W. Reynders, Los Alamos National Laboratory, USA
Tony Skjellum, Mississippi State University, USA

Table of Contents

Runtime Support for Multi-tier Programming of Block-Structured Applications on SMP Clusters*

Stephen J. Fink and Scott B. Baden

Department of Computer Science and Engineering
University of California, San Diego
La Jolla, CA 92093-0114

Abstract. We present a small set of programming abstractions to simplify efficient implementations for block-structured scientific calculations on SMP clusters. We have implemented these abstractions in KeLP 2.0, a C++ class library. KeLP 2.0 provides hierarchical SMPD control flow to manage two levels of parallelism and locality. Additionally, to tolerate slow inter-node communication costs, KeLP 2.0 combines inspector/executor communication analysis with overlap of communication and computation. We illustrate how these programming abstractions hide the low-level details of thread management, scheduling, synchronization, and message-passing, but allow the programmer to express efficient algorithms with intuitive geometric primitives.

1 Introduction

Multi-tier parallel computers, such as clusters of symmetric multiprocessors (SMPs), have emerged as important platforms for high-performance computing [1]. A multi-tier computer, with several levels of locality and parallelism, presents a more complex non-uniform memory hierarchy than a single-tier multicomputer with uniprocessor nodes. In order to use multi-tier platforms efficiently, the programmer or compiler must orchestrate parallelism and locality to match the hardware capabilities.

On single-tier parallel computers, MPI [2] and HPF [3] have emerged as standard approaches to portable parallel programming. However, the proper programming model for multi-tier parallel computers remains an unresolved issue. At present, the programmer faces myriad options regarding the coordination of heavyweight processes, lightweight threads, shared memory, message-passing, synchronization, scheduling, and load balancing [4]. This daunting array of low-level programming detail hinders efficient implementations for multi-tier platforms.

---

* Stephen Fink was supported by the DOE Computational Science Graduate Fellowship Program, Scott Baden by NSF contract ASC-9520372. Computer time on the Maryland Digital AlphaServer was provided by NSF CISE Institutional Infrastructure Award CDA9401151 and a grant from Digital Equipment Corp.. The authors wish to thank Joel Saltz and Alan Sussman for arranging access to the AlphaServer.

We present a small set of programming abstractions to simplify implementation of efficient algorithms for block-structured scientific calculations on SMP clusters. This paper extends previous work [5] with two contributions specifically targeted for multi-tier architectures: hierarchical SPMD control flow, and overlap of communication and computation. We show how high level abstractions hide tedious low-level implementation details, but allow the programmer to express efficient algorithms with intuitive geometric primitives.

We have implemented these abstractions in KeLP 2.0, a C++ class library running on a Digital AlphaServer 2100 cluster. Performance results on three codes show that multi-tier programming improves performance substantially compared to codes that rely on straightforward MPI. Furthermore, the results illustrate the benefits of overlapping communication and computation to tolerate inter-node message-passing costs.

2 Programming Abstractions

2.1 Structural Abstraction

The KeLP programming abstractions extend *structural abstraction*, a programming model introduced in the LPARX programming system [6]. Under structural abstraction, first-class meta-data objects represent the geometric structure of a calculation. Previous work describes KeLP abstractions to manage irregular block data decompositions and communication on single-tier multicomputers [5].

To discuss multi-tier KeLP programming, we first introduce the following meta-data abstractions: *Region*, *Map*, *FloorPlan*, and *MotionPlan*. KeLP 2.0 implements each of these abstractions as a first-class C++ object.

The Region represents a rectangular subset of Z^n; i.e., a regular section with stride one. KeLP provides the *Region calculus*, a set of high-level geometric operations to help the programmer manipulate Regions. Typical Region calculus operations include *shift*, *intersection*, and *grow*.

The Map class implements a function $Map : \{0, \ldots, k-1\} \to Z$, for some integer k. That is, for $0 \le i < k$, $Map(i)$ returns some integer. The Map forms the basis for node and processor assignments in KeLP partitioning.

The FloorPlan consists of a Map along with an Array of Regions. The FloorPlan can represent a potentially irregular block data decomposition. Alternatively, a FloorPlan can represent distribution of work among processors of a single SMP.

The MotionPlan implements a first-class, user-level block communication schedule [7]. The programmer builds and manipulates MotionPlans using geometric Region calculus operations.

2.2 Hierarchical Control Flow

The multi-tier KeLP abstractions support three levels of control: a *collective* level, a *node* level, and a *processor* level. The collective level manages data distribution and communication among SMP nodes. The node level manages partitioning and communication among the multiple processors at a single SMP node. The processor level executes a serial instruction stream on a single physical processor.

The KeLP program starts in the collective level, and descends to the node level and to the processor level through two iterators: the nodeIterator and procIterator. Given a Map M, a nodeIterator executes the ith loop iteration on node $M(i)$. Thus, each loop iteration forms an independent node-level instruction stream. From the node-level stream, the program descends to the processor level via a procIterator. Each procIterator iteration executes serially a single processor. At the processor level, the programmer may invoke numerical kernels in an extrinsic sequential language such as Fortran.

2.3 Storage Model

KeLP objects represent two kinds of data: meta-data and Grid data. The meta-data, as represented by Regions, FloorPlans, etc., lives at all three levels of the control flow. Each program level manipulates meta-data to orchestrate application structure at lower program levels.

The basic unit of actual data is the Grid. A Grid is an array of objects of some type T, whose index space is a Region. For example, the Fortran array real A(3:7) corresponds to a one-dimensional Grid A of real with *region(A)* = [3,7]. A Grid lives in exactly one node's address space; a single Grid is not distributed across multiple nodes. The program can access the Grid data from the node-level instruction stream at that node, or from processor-level instruction streams nested at that node.

An XArray is an array of Grids, whose structure is represented by a Floor-Plan. For an XArray X, $X(i)$ denotes the ith Grid in X.

2.4 Data Motion

KeLP performs inspector/executor analysis [7] with the MotionPlan and Mover classes. The Mover, a first-class executor object, performs the data motion represented by a MotionPlan as an atomic collective operation. For a more complete discussion, see [5].

To achieve good performance on SMP clusters, it is vital to tolerate slow inter-node communication delays. To facilitate overlap of communication and computation, multi-tier KeLP Movers provide asynchronous execution. For a Mover M, M.start() begins asynchronous execution of a data motion pattern. M.wait() blocks until the data motion pattern completes. Individual nodes or processors may block on M.wait(), while other nodes or processors continue execution. Thus, the programmer may selectively block individual nodes or processors as dictated by the data dependencies of the application.

Multi-tier KeLP also implements barriers, reductions, and broadcasts across and within SMP nodes.

3 Programming example

To illustrate KeLP's multi-tier programming constructs, we briefly discuss the implementation of redblack3D, a 7-point stencil relaxation to solve Poisson's equation over a cube.

The usual SPMD implementation employs a BLOCK data decomposition and carries additional ghost cells to buffer off-processor data. Each relaxation consists of two steps: (1) communicate with nearest neighbors to exchange ghost cell values, and (2) independently relax on the local portion of the global mesh.

Fig. 1 shows multi-tier KeLP code to implement the relaxation. The Relax subroutine of Fig. 1 starts in collective control flow. At the collective level, the program invokes a Mover object to interpret the communication pattern and exchange ghost cells between SMP nodes (lines 2-3).

At line 4, the program drops to node level control via the nodeIterator. At line 5, the program queries the nodeIterator to determine the number of the current iteration. Each iteration of the loop executes on exactly one SMP node according to the Map associated with XArray X; iteration n executes on the node that owns Grid X(n). At the node level, we partition the work at SMP node n by setting up a FloorPlan F, describing the spatial work decomposition for node n (line 6). IntraU(), a user-written routine, generates the FloorPlan F, uniformly partitioning the domain of Grid X(n). Typically, the programmer will employ a library of common partitioning routines.

At line 7, the program drops to processor level control via the procIterator. The FloorPlan F describes the intra-node work partitioning for node n. In particular, processor p relaxes on Grid X(n) over the Region F(p). Line 9 invokes a serial numeric kernel, possibly in an extrinsic language such as Fortran 77, to perform the relaxation.

```
(1) Relax(XArray X, XArray rhs,
        Mover Mvr, int rb) {
(2)  Mvr.start();
(3)  Mvr.wait();
(4)  for (nodeIterator ni(X); ni; ++ni) {
(5)   int n = ni();
(6)   FloorPlan F = intraU(X(n));
(7)   for (procIterator pi(F); pi; ++pi) {
(8)    int p = pi();
(9)    serialRelax(X(n),rhs(n),F(p),rb);
(10) } } }
```

Fig. 1. Redblack3D relaxation with no overlap.

We can improve performance by overlapping communication and computation, as shown in Fig. 2. At line 2, the Mover asynchronously starts communication. Instead of waiting for the Mover as a collective operation, we immediately begin relaxation on an interior region of the grid which does not depend on off-node ghost values.

Fig. 2b illustrates the phenomenon in 2D for a two-processor SMP. The two interior Regions marked 0 and 1 do not depend on incoming ghost cell values. The surrounding annulus, Regions 2 through 5, must wait for incoming ghost cell values (shaded). Line 5 calls `intraUDep()`, a user-written routine to generate a FloorPlan F describing this intra-node partitioning.

With FloorPlan F set up, the procIterator loop in Fig. 2a proceeds as follows. The first P (in the example, $P = 2$) domains do not wait on the Mover, and execute immediately, in parallel. However, the remaining domains depend on ghost cell values, and must wait until communication completes. The Mover `wait()` call at line (8) enforces this constraint.

```
(1)  Relax(XArray X, XArray rhs,
        Mover Mvr, int rb) {
(2)    Mvr.start();
(3)    for (nodeIterator ni(X); ni; ++ni) {
(4)      int n = ni();
(5)      FloorPlan F = intraUDep(X(n));
(6)      for (procIterator pi(F); pi; ++pi) {
(7)        int p = pi();
(8)        if (p >= P) Mvr.wait();
(9)        serialRelax(X(n),rhs(n),F(p),rb);
(10) } } }
```

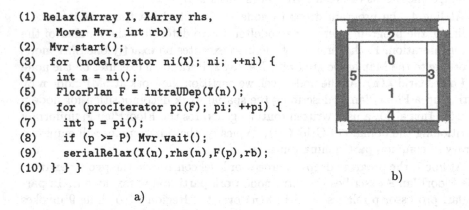

a)

b)

Fig. 2. a) Redblack3D relaxation with overlap. b) Intra-node FloorPlan isolates off-node data dependencies for a 2-processor SMP node.

4 Performance Results

We have implemented the multi-tier KeLP infrastructure as a C++ class library on a cluster of Digital AlphaServer 2100's running Digital UNIX 4.0. Each SMP has four Alpha 21064A processors, and each processor has a 4MB direct-mapped L2 cache. For inter-node communication, we rely on MPICH 1.0.12 [8] over an OC-3 ATM switch. Using a simple ring test we observe a message start time of 745 μs and a peak bandwidth of 12 MB/sec.

Unfortunately, we encountered severe problems with Digital UNIX 4.0 scheduling of lightweight threads. The operating system provides no user control over lightweight thread processor assignment, and we could not reliably utilize more than two processors with the POSIX thread package. As a result, we were forced to emulate "threads" using heavyweight processes with a memory-mapped shared heap. In our implementation, the extra overhead for spawning heavyweight processes occurs only at program startup.

To implement overlap of communication overhead at an SMP node, we spawn an extra thread devoted entirely to communication. We experimented with various options for scheduling the multiple threads onto the processors at an SMP node. We found that the best performance results when the implementation explicitly binds threads to processors.

We now report performance results for three applications in KeLP 2.0. For each application, we compare a straightforward MPI code to multi-tier KeLP code, with and without overlap of communication with computation.

The first application, redblack3D, performs a 7-point relaxation to solve Poisson's equation over a unit cube as described in Section 3. The MPI version of this code uses BLOCK data decomposition on all three axes.

The second application, SUMMA, implements dense matrix multiplication using the SUMMA algorithm [9]. The MPI code for SUMMA is listed in [9] and was made publicly available by the authors. The serial matrix multiply kernel calls vendor-provided BLAS. The multi-tier KeLP code for SUMMA uses a two-level decomposition of the algorithm. The outer level uses the SUMMA algorithm between SMP nodes. Within each SMP node, we parallelize the matrix multiplication with a simple domain decomposition. To overlap communication and computation, we developed a multi-tier pipelined version of the SUMMA algorithm, which will be described in detail elsewhere [10].

The third application is the NAS-FT benchmark, which solves a 3D diffusion equation using Fast Fourier Transform. We obtained MPI code for FT from the NAS Parallel Benchmarks v2.1 [11]. The multi-tier KeLP versions add a second level of parallelism to node-level kernels with domain decomposition. To overlap communication and computation, we pipeline the FFTs across iterations with the algorithm described in [12].

Fig. 3 reports performance of these codes, scaling the problem size with the number of nodes. The results show that on eight SMP nodes, the multi-tier KeLP code without overlap outperforms the MPI code by a factor of 2 for redblack3D and a factor of 4.6 for NAS-FT. These wide discrepancies reflect an inefficient MPI implementation, which does not exploit shared memory hardware for intra-node messages. The results also show that explicit overlap of communication with computation further improves performance by 22% and 13% on these two codes.

The SUMMA MPI code uses a pipelined algorithm to overlap communciation between processors. On eight nodes, the MPI code's pipelining strategy outperforms multi-tier KeLP code without overlap. The multi-tier KeLP code with overlap improves performance by 33% with a multi-tier pipelined algorithm.

For all three codes, the results show that the benefits of communication overlap increase with the number of nodes.

5 Conclusion and Related Work

We have presented high-level programming abstractions for block-structured scientific calculations on multi-tier parallel computers. The abstractions allow the

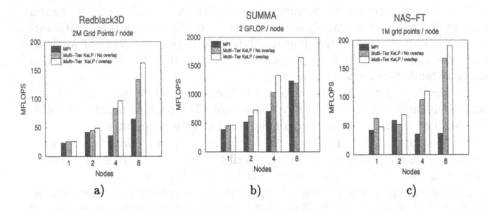

Fig. 3. Performance of a) redblack3D, b) SUMMA, and c) NAS-FT.

programmer to express efficient algorithms with intuitive geometric operations, while hiding the low-level details of message-passing, synchronization, thread management, and scheduling. Performance results on three codes indicate that the multi-tier approach outperforms pure message-passing codes. Moreover, the results show that overlap of communication and computation effectively improves performance.

The KeLP programming abstractions extend the structual abstraction model introduced in the LPARX programming system [6]. KeLP's communication model combines structural abstraction with inspector/executor communication analysis as introduced in Multiblock PARTI[7].

In the Phase Abstractions programming model, Snyder [13] advocated separation of programs into levels corresponding to collective(Y) and node(X) levels. Control flow in multi-tier KeLP programming model extends Snyder's XYZ program levels to multi-tier machines. Alpern, Carter, and Ferrante's PMH model [14] provides an elegant framework for multi-tier parallel architectures.

The Cedar Fortran language [15] was perhaps the first language to incorporate two levels of parallelism in order to match a hierarchical parallel architecture. Some more recent systems explicitly target SMP clusters. Sawdey and O'Keefe [16] have applied the Fortran-P programming model to grid-based applications on SMP clusters. In Fortran-P, a compiler translates serial grid-based code to explicitly threaded parallel code. KeLP 2.0 presents a more complex, explicitly parallel model, but allows the programmer to express a wider class of algorithms and exert more control over the implementation.

Bader and JáJá have developed SIMPLE [17], a set of collective communication operations for SMP clusters. SIMPLE provides more general, lower-level primitives than KeLP 2.0, and does not help with data decomposition or overlap of communication and computation. The KeLP 2.0 implementation relies on collective operations similar to those provided by SIMPLE.

In future work, we will consider performance models for multi-tier KeLP applications, and investigate multi-tier programming for irregular block-structured applications.

References

1. P. R. Woodward, "Perspectives on supercomputing: Three decades of change," *IEEE Computer*, vol. 29, pp. 99–111, October 1996.
2. Message-Passing Interface Standard, "MPI: A message-passing interface standard," University of Tennessee, Knoxville, TN, Jun. 1995.
3. High Performance Fortran Forum, *High Performance Fortran Language Specification*, Nov. 1994.
4. W. W. Gropp and E. L. Lusk, "A taxonomy of programming models for symmetric multiprocessors and SMP clusters," in *Proceedings 1995: Programming models for massively parallel computers*, pp. 2–7, October 1995.
5. S. J. Fink, S. B. Baden, and S. R. Kohn, "Flexible communication mechanisms for dynamic structured applications," in *Proc. 3rd Int'l Workshop IRREGULAR '96*, pp. 203–215, Aug. 1996.
6. S. R. Kohn, *A Parallel Software Infrastructure for Dynamic Block-Irregular Scientific Calculations*. PhD thesis, University of CA at San Diego, 1995.
7. G. Agrawal, A. Sussman, and J. Saltz, "An integrated runtime and compile-time approach for parallelizing structured and block structured applications," *IEEE Transactions on Parallel and Distributed Systems*, vol. 6, Jul. 1995.
8. W. Gropp, E. Lusk, N. Doss, and A. Skjellum, "A high-performance, portable implementation of the MPI message passing interface standard," tech. rep., Argonne National Laboratory, Argonne, IL, 1997. http://www.mcs.anl.gov/mpi/mpich/.
9. R. van de Geign and J. Watts, "SUMMA: Scalable universal matrix multiplication algorithm," *Concurrency: Practice and Experience*, vol. 9, pp. 255–74, April 1997.
10. S. J. Fink, "Hierarchical programming for block–structured scientific calculations." In preparation.
11. D. Bailey, T. Harris, W. Saphir, R. van der Wijngaart, A. Woo, and M. Yarrow, "The NAS parallel benchmarks 2.0," Tech. Rep. NAS-95-020, NASA Ames Research Center, December 1995.
12. R. C. Agarwal, F. G. Gustavson, and M. Zubair, "An efficient parallel algorithm for the 3-d FFT NAS parallel benchmark," in *Proc. of SHPCC '94*, pp. 129–133, May 1994.
13. L. Snyder, "Foundations of practical parallel programming languages," in *Portability and Performance of Parallel Processing* (T. Hey and J. Ferrante, eds.), John Wiley and Sons, 1993.
14. B. Alpern, L. Carter, and J. Ferrante, "Modeling parallel computers as memory hierarchies," in *Programming Models for Massively Parallel Computers* (W. K. Giloi, S. Jahnichen, and B. D. Shriver, eds.), pp. 116–23, IEEE Computer Society Press, Sept. 1993.
15. R. Eigenmann, J. Hoeflinger, G. Jaxson, and D. Padua, "Cedar Fortran and its compiler," in *CONPAR 90-VAPP IV. Joint International Conference on Vector and Parallel Parocessing*, pp. 288 – 99, 1990.
16. A. C. Sawdey, M. T. O'Keefe, and W. B. Jones, "A general programming model for developing scalable ocean circulation applications," in *Proceedings of the ECMWF Workshop on the Use of Parallel Processors in Meteorology*, January 1997.
17. D. A. Bader and J. JáJá, "SIMPLE: A methodology for programming high performance algorithms on clusters of symmetric multiprocessors." Preliminary Version, http://www.umiacs.umd.edu/research/EXPAR/papers/3798.html.

Automatic Run-Time Code Generation in C++

Nobuhisa Fujinami

Sony Computer Science Laboratory Inc.

Abstract. Run-time code generation (RTCG) enables program optimizations specific to values that are unknown until run time and improves the performance. This paper shows that object-oriented languages allow RTCG from a simple analysis of source programs. It also proposes an automatic, safe, and efficient RTCG system implemented as a preprocessor of a C++ compiler. It can inline virtual member functions combined with other optimizations. Increases in speed have ranged from 1.2 to 2.3 times.

1 Introduction

Run-time code generation (RTCG) is a kind of partial evaluation performed at run time. It generates machine code specific to values that are unknown until run time and enhances the speed of a program, while preserving its generality. RTCG itself is becoming a mature technique. Description languages [1] and systems [2] for run-time code generators have been proposed. Systems that automatically generate run-time code generators from source programs [3, 4] have also been proposed, but they do not automate when the code is generated or how it is used. The programmer must instead rewrite the source code to initiate RTCG, to invoke the generated machine code, and to maintain consistency between the embedded values in the code and the actual values.

This paper shows that object-oriented (O-O) languages provide useful information for RTCG. It also proposes a system for automatic RTCG, generated code invocation, and code management. The author has already proposed an automatic RTCG system for C++ that generates machine code when objects are instantiated [5, 6]. This paper extends that system and proposes a more general method.

The goals of the RTCG system are twofold. The first is *automation*—minimal rewriting of source programs is required to use RTCG. The programmer is freed from annotating programs and from providing suitable parameters to preserve consistency. This goal is achieved by using an O-O language as the source language. Some methods of objects are replaced with machine code routines generated at run time. The values of some instance variables are embedded into the routines with optimizations specific to the values. The programmer needs only to specify the methods to be optimized and to invoke the system, which automatically generates run-time code generators and inserts code for invoking them, and for invoking and invalidating the generated code.

The second goal is *efficiency*—the run-time code generator and the code it generates should be optimized to increase the cases where RTCG is profitable. This goal is similar to existing systems, but the author's system uses a more aggressive method. Not only does it generate specialized code generators for the methods, like existing systems, but it also embeds specialized optimization routines, such as global run-time constant propagators specific to the program, into the code generators.

2 Rationale for O-O Languages in RTCG

RTCG improves the efficiency of programs by generating machine code optimized to input values or intermediate results that are unknown until run time. If programs are written in O-O languages, it is natural to define objects with instance variables that represent such values. An example is the generation and rendering of a 3-D scene. The programmer may represent the scene, a run-time constant during rendering, as an object with instance variables representing a set of graphics objects, a viewing point, and light sources. The scene object's rendering methods can be optimized through RTCG.

The benefits of focusing on instance variables of objects are twofold. First is the automation of the timing of code generation/invalidation. Because of encapsulation, all assignments to non-public instance variables (`private` data members in C++) can be known, except for indirect accesses through pointers, from the definition of the class and its methods. Values of these variables can be embedded consistently into generated machine code.

Second is the automation of managing generated code. Since generated code can be viewed as a part of the object, its management can be left to the object construction/destruction mechanism of O-O languages. Code is discarded when the corresponding object is deleted. Management of several code versions for the same method is trivial.

The author's system requires the programmer to mark the methods to which RTCG will be applied. Automatic detection of the applicability is possible but not practical, because a too-aggressive application of RTCG increases compilation time and executable size.

The implementation uses C++ as its input language. Keyword `runtime` before a declaration of a member function marks it for RTCG. The system assumes that all the "known" data members (see Subsection 3.2) used but not changed in that member function are run-time constants. If some data members are modified frequently, and the programmer does not want their values to be embedded into the generated code, the programmer can put keyword `dynamic` before the definitions of the members.[1]

Figure 1 shows an example of a class that represents a set of graphics objects. Class `objectTableType` has two `private` data members (`count` and `table`), a constructor, and two `public` member functions (`add` and `intersect_all`). If

---

[1] Inappropriate insertions of keywords `runtime` and `dynamic` may degrade performance, but they cannot change the meaning of the program.

```
class objectTableType {
  int count;                        // graphics object counter
  objectType* table[MAXOBJECT];     // pointers to graphics objects
public:
  objectTableType(): count(0) {}
  int add(objectType *p)            // add graphics object
  const objectType* intersect_all(rayType &, float &);
};                                  // return the first object the ray intersects
```

Fig. 1. Definition of class `objectTableType`.

the programmer inserts `runtime` before the declaration of `intersect_all` to request RTCG, because it is executed many times during rendering, the system then generates machine code for `intersect_all`, assuming data members `count` and `table` are run-time constants. An invocation of member function `add`, which modifies the data members, invalidates the generated code.

3 Implementation

The system is designed as a preprocessor of a C++ compiler. The current implementation is for a Borland C++ compiler running on 80x86 based computers with a Win32 API. The preprocessor is written in C++ (about 11,000 lines of source code), and the executable file name is RPCC.EXE. Its usage is very simple. If a source program is in one file and a normal command line for compilation is `bcc32 -O2 file.cpp`, then the following command lines:

`cpp32 file.cpp`   Process C++ preprocessor directives, create `file.i`
`rpcc file.i`   Emit C++ source including run-time code generators
`bcc32 -O2 -P file.out` Compile the output

enable RTCG. The technique used in the pre-pass in [7] will allow separate compilation, but this is not currently supported.

3.1 Organization of the System

Figure 2 shows an overall organization of the implementation. Note the distinction between compile time and run time. At compile time, C++ preprocessor directives in a source program are processed first. Then RPCC.EXE rewrites the program to embed code generators, to insert code for invoking them and for invoking and invalidating the generated code. The output is compiled into an executable file. The source program and its intermediate representation are manipulated only at this compile time.

At run time, code generators are invoked with run-time constants as parameters. Each code generator is specific to one member function. One code generator may generate several machine code routines with different values of run-time constants. The generated routines, which are expected to be more efficient than statically compiled ones, are invoked if necessary.

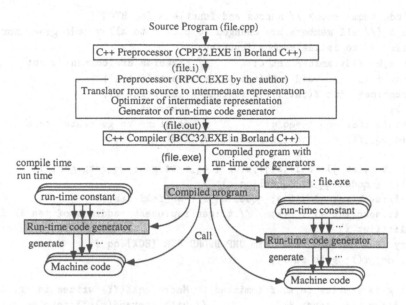

Fig. 2. Organization of the implemented system.

Figure 3 shows an example of an input to RPCC.EXE. Figure 4 shows the output (comments are added for readability). RPCC.EXE processes member functions with keyword `runtime` and generates run-time code generators in C++. Pointers to generated machine code routines are added to a class as data members, and code generators are added as member functions. The processed member functions are replaced with code fragments to invoke the code generators if necessary, and to invoke the generated code. RPCC.EXE also inserts code for deleting generated machine code in the destructors and in the member functions that modify the data members embedded in the generated machine code.

Since the output code generators are in C++, most of the optimizations of the code generators themselves (inlining code generation routines, constant folding) are left to the C++ compiler. This also simplifies the implementation of the generator of code generators.

```
class A {                          runtime int f(int x);
 private:                          runtime int g(int x);
  int a,b;                        };
  dynamic int c;                  A::A(int i,int j): a(i),b(j),c(0) {}
 public:                          void A::setb(int j) { b=j; }
  A(int i,int j);                 void A::setc(int k) { c=k; }
  void setb(int j);               int  A::f(int x) { return a*b-x; }
  void setc(int k);               int  A::g(int x) { return a*c-x; }
```

Fig. 3. Example of class definition and implementation.

```
#include <qqmacro.h> // macros and functions for RTCG
class A {// all members are changed to public to allow code generators
 public: // to inline member functions
   int a,b; /*dynamic*/ int c;        // keywords are commented out
   A(int i,int j); void setb(int j); void setc(int k);
   /*runtime*/ int f(int x); /*runtime*/ int g(int x);
   ~A();                              // destructor
   mutable char *qq_f,*qq_g;          // pointers to generated code
   void qq__f() const,qq__g() const; // code generators
};
A::A(int i,int j): a(i),b(j),c(0) { qq_f=&f:gen; qq_g=&g:gen; }
A::~A() { qqdel(qq_f); qqdel(qq_g); }
void A::setb(int j) { b=j; qqdel(qq_f); qq_f=&f:gen; }
void A::setc(int k) { c=k; } // &f:gen represents address of gen in f
int A::f(int x) {
 retry: asm MOV ECX,this; asm JMP DWORD PTR [ECX].qq_f;
  gen: qq__f(); goto retry;
}
// A::g is similar to A::f (omitted).  Macro qqXX(YY) writes instr. XX
void A::qq__f() const {              // with operand(s) YY into memory
  // prologue code generator (omitted)
  qqMOVi(6,a*b);                     // MOV ESI,a*b
  qqSUBdx(6,5,12);                   // SUB ESI,[EBP+12] ; x
  qqMOV(0,6);                        // MOV EAX,ESI
  // epilogue code generator (omitted)
}
void A::qq__g() const {
  // prologue code generator (omitted)
  MOVdx(7,5,8);                      // MOV EDI,[EBP+8] ; this
  if(a!=0) { // algebraic simplification
    qqMOVdx(6,7,offsetof(A,c)); // MOV ESI,[EDI].c
    qqMUL_I(6,a);                    // IMUL ESI,a (strength may be reduced)
  } else {
    qqMOV_I(6,0);                    // XOR ESI,ESI
  }
  qqSUBdx(6,5,12);                   // SUB ESI,[EBP+12] ; x
  qqMOV(0,6);                        // MOV EAX,ESI
  // epilogue code generator (omitted)
}
```

Fig. 4. Example of output from RPCC.EXE.

RPCC.EXE consists of three phases like conventional compilers (see Fig. 2). The intermediate representation format used in these phases are designed to be suitable for generating run-time code generators. Expressions (except conditional operators and function invocations) are represented as directed acyclic graphs (DAGs) that preserve the semantics of C++ expressions. Simple traversal of DAGs can classify language constructs into compile time, code generation time, or execution time, using "known" marks in the symbol table.

3.2 Program Analysis

Assume that all the member functions of a class and its derived classes are compiled together. Static members are not considered in the analysis. In the first phase, all private, protected, or const[2] data members of the class without keyword dynamic are marked "known". Then, the mark for the non-const data member is cleared, if its address is taken in the member function, if the address is passed directly or via casts, + or − to a variable, as a parameter, or as a return value, and if the type of the destination is not a pointer/reference to a const. The values of the data members still marked "known" are explicitly used or modified only by the member functions of the class.

Let F be a member function with keyword runtime, and let X be any data member marked "known" after the process above. If X is used but not changed in F, X is treated as a run-time constant in the code generator for F. Then if another member function G changes X, code to invalide run-time machine code for F is inserted into G (see Fig. 3 and 4).

Automatic specialization systems for rich languages like C [3, 8] require sophisticated alias analysis because C does not support encapsulation well. In the author's experience, the simple method above suffices to analyze programs for RTCG. It does not force the programmer to use unnatural consts or privates.

3.3 Optimization

Among the optimizations of the machine code generated by the run-time code generator, those detected at compile time are treated in a way similar to conventional code optimizations. Because those detected at code generation time (i.e. run time) do not use any intermediate representation of the source program, the optimizer should be hard-coded into the run-time code generator. RPCC.EXE embeds code for run-time optimizations (RTOs), such as local constant propagation/folding, strength reduction, redundant code elimination, and algebraic simplification where necessary. It also performs non-trivial RTOs, such as global constant propagation, complete loop unrolling, and inlining virtual functions. It can inline, with the help of other RTOs, function calls that conventional static analysis method cannot detect.

4 Evaluation

Evaluation of the implementation used a NEC PC-9821La10/8 (P5-100), the compiler is Borland C++ 4.52J, and the compiler options are "-5 -O2" (Pentium, optimize for speed).

The first test is a single-precision, floating-point matrix multiply implemented as inner products of class Vector. If matrices are sparse, there is a room for RTO. Figure 5 shows execution times of two cases. In the left graph, 90% of the ele-

---

[2] It may violate the assumption of the analysis to cast a pointer to const into a pointer to non-const. Such an attempt is considered to be illegal because it is not safe to modify an object through such a pointer.

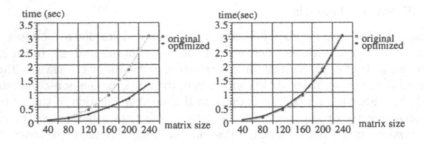

Fig. 5. Sparse and dense matrix multiplication.

ments are zero. All elements are determined at run time. The "original" program is compiled by Borland C++ alone, and the "optimized" one using RPCC.EXE and Borland C++, both from the same source program. The "optimized" one is faster if matrices are large enough. If the matrix size is 240 × 240, speedup is about 2.3. The break-even point is about 60 × 60. This shows fairly low code generation cost. Since the code fragment for a non-zero element in the "optimized" program is similar to that for each element in the "original" program, speedup is mostly due to RTCG. In the generated code, the dominant optimization is run-time constant folding of zeros combined with loop unrolling. In the right graph of Fig. 5, the matrices are dense. The "optimized" program is slightly slower than the "original," due to the code generation cost.

Table 1 shows additional results. The first program, a ray tracer, reads a scene file at run time and displays the ray-traced image (512 × 512 full color image in the evaluation). Speedup is due to determination of virtual member function invocation combined with inlining. The second program is a box-packing puzzle solver. It reads a puzzle definition file that describes the box and the pieces and prints the solutions. RTCG is applied to member function put of class Piece. Its instance variables are the shapes of the pieces, and are run-time constants. Speedup is due to loop unrolling and constant folding. The third program is a numerical integrator of a differential equation whose parameters are given at run time. Speedup is due to run-time constant folding.

5 Conclusion and Future Work

This paper describes an automatic RTCG, generated code invocation, and code management system for O-O languages such as C++. The characteristics of the proposed system include: little effort required from the programmer; system

Table 1. Further evaluation results.

Program	Exec Time (org/opt)	Ratio	Comment
Ray tracer	42.6 / 30.4 sec	1.40	P6-150MHz+Millennium
Puzzle solver	2.72 / 2.28 sec	1.19	-
Numerical integrator	30.67 / 21.50 sec	1.43	-

safety in the face of inappropriately inserted RTCG keywords; efficient run-time code generators; and efficient output from these generators. The implementation and the evaluation results show that an O-O language allows RTCG from a simple analysis of source programs.

Most efforts to improve the performance of O-O programming systems have been in optimizations like imperative programming systems. O-O, however, provides greater freedom in optimizations by compilers because the implementation of an object is separated from its interface. This paper has shown that optimizations through RTCG are applied naturally to programs written in O-O languages.

Future work will extend the system to optimize commonly used groups of objects, such as linked lists and hash tables. Operations on these data structures can be represented as code fragments held in the objects. This is similar to *executable data structures* or *Quajects* by Massalin [9, 10], which were implemented using handwritten templates in an assembly language. O-O languages may permit automatic application of this optimization.

Acknowledgments I express my gratitude to Dr. Mario Tokoro for supervising the research. I also appreciate many suggestions by Dr. Satoshi Matsuoka and Dr. Calton Pu as well as the work of programming ray tracer in C++ by Ms. Kayoko Sakai. Finally, I would like to thank the members of Sony CSL for their valuable advice.

References

1. Dawson R. Engler, Wilson C. Hsieh, and M. Frans Kaashoek. 'C: A language For High-Level, Efficient, and Machine-independent Dynamic Code Generation. In *Conference Record of POPL '96*, pp. 258–270, January 1996.
2. Joel Auslander, Matthai Philipose, Crig Chambers, Susan J. Eggers, and Brian N. Bershad. Fast, Effective Dynamic Compilation. In *Proceedings of the SIGPLAN '96 Conference on PLDI*, pp. 149–159, May 1996.
3. Charles Consel, Luke Hornof, François Nöel, and Nicolae Volanshi. A Uniform Approach for Compile-time and Run-time Specialization. Technical Report No. 2775, INRIA, January 1996.
4. Mark Leone and Peter Lee. A Declarative Approach to Run-Time Code Generation. In *Workshop Record of WCSSS'96*, pp. 8–17, February 1996.
5. Nobuhisa Fujinami. Run-Time Optimization in Object-Oriented Languages. In *Proceedings of 12th Conference of JSSST*, September 1995. In Japanese. Received Takahashi Award.
6. Nobuhisa Fujinami. Automatic and Efficient Run-Time Code Generation Using Object-Oriented Languages. *Submitted to Computer Software, JSSST*, 1996. In Japanese.
7. Gerald Aigner and Urs Hölzle. Eliminating Virtual Function Calls in C++ Programs. In *Proceedings of ECOOP'96*, June 1996.
8. Lars Ole Andersen. *Program Analysis and Specialization for the C Programming Language.* PhD thesis, DIKU, University of Copenhagen, May 1994.
9. Calton Pu, Henry Massalin, and John Ioannidis. The Synthesis kernel. *Computing Systems*, Vol. 1, No. 1, pp. 11–32, Winter 1988.
10. Henry Massalin. *Synthesis: An Efficient Implementation of Fundamental Operating System Services.* PhD thesis, Graduate School of Arts and Sciences, Columbia University, April 1992.

A Comparison of Performance-Enhancing Strategies for Parallel Numerical Object-Oriented Frameworks

Federico Bassetti, Kei Davis, and Dan Quinlan

Los Alamos National Laboratory

Abstract. Performance short of that of C or FORTRAN 77 is a significant obstacle to general acceptance of object-oriented C++ frameworks in high-performance parallel scientific computing; nonetheless, their value in simplifying complex computations is inarguable. Examples of good performance for object-oriented libraries/frameworks are interesting, but a systematic analysis of performance issues has not been done. This paper explores a few of these issues and reports on three mechanisms for enhancing the performance of object-oriented frameworks for numerical computation. The first is binary operator overloading implemented with substantial internal optimizations, the second is expression templates, and the third is an optimizing preprocessor. The first two have been completely implemented and are available in the A++/P++ array class library[1], the third, ROSE++[2], is work in progress. This paper provides some perspective on the types of optimizations that we consider important in our numerical applications using OVERTURE[3] involving complex geometry and AMR on parallel architectures.

1 Introduction

The use of object-oriented C++ frameworks has greatly simplified the development of complex serial and parallel scientific applications at Los Alamos National Laboratory (LANL) and elsewhere. Examples from LANL include OVERTURE [1] which supports complex geometries, adaptive mesh refinement (AMR), and overlapping grid computations as well as simpler single rectangular grid computations, and POOMA [8] which has an emphasis on support for particle and molecular dynamics. In spite of considerable use of and commitment to these frameworks, concerns about performance are a significant issue; performance very close to that of carefully hand-crafted C or FORTRAN 77 with message passing must be achieved before acceptance and use of such frameworks will be widespread. Our goal is to characterize aspects of C++ usage that give poorer performance than C and to provide mechanisms to minimize or eliminate the performance penalties; we believe this goal to be fully realizable.

---

[1] A++/P++ is available from http://www.c3.lanl.gov/cic19/teams/napc/napc.shtml
[2] ROSE++ Web Site: http://www.c3.lanl.gov/ dquinlan/ROSE.html
[3] OVERTURE is available from http://www.c3.lanl.gov/cic19/teams/napc/napc.shtml

The potential performance of three execution models for implementing such frameworks is explored: implementation with *overloaded binary operators*, implementation with *expression templates (ETs)*, and *semantics-based source-to-source transformation* (a new approach) with a special-purpose preprocessor. Naive implementation of any of these techniques will suffer putative disadvantages, so in the context of our comparison we have focused on the practical limitations of each.

For numerical applications array types are basic, and the performance of their implementation has a fundamental impact on the performance of libraries and applications built on them. Our target for optimization is the parallel distributed dynamic array class A++/P++, a component of OVERTURE, which incorporates normal array operations, indirect addressing, and block "where" statements. We report our experiences with A++/P++ using each of the three techniques. A++/P++ has been implemented both using overloaded binary operators and ETs. In some cases the data for ETs is from hand-written idealizations of ET expansion; this is fair for determining an upper bound on ET performance while simplifying the gathering of performance data. Similarly, since the preprocessor is still in development we compare directly to C code that the current preprocessor can or can be expected to produce; however, we make no specific claims about the preprocessor. Instead, we concentrate on issues in achieving C performance from C++ applications using array-like constructs.

For each technique two optimizations are evaluated in terms of performance at the top of the memory hierarchy: avoidance of register spillage and efficient use of cache. Other issues are scheduled for future work.

2 Execution Models

Detailed discussions of operator overloading and ETs may be found in the literature [4–7]. The third technique that we are developing is semantics-based source-to-source transformation. This technique performs optimizations much like ETs do without their concomitant defects, and also applies optimizations such as loop fusion and rescheduling and aggregating communications. This technique is *semantics based* because its implementation has exact knowledge of the semantics of the classes over which it optimizes. It is *source-to-source* because both its input and output are C++ code. This built-in knowledge of class semantics makes possible optimizations that could not in general be performed automatically by a compiler. Further detail exists elsewhere [2].

3 Issues in the Optimization of Numerical Applications

We address only two of the many issues in optimizing numerical applications here: minimization of register spillage, and loop fusion. Other issues are under investigation.

3.1 Register Spillage

Register spillage occurs when a program needs more registers than are available, resulting in register values being stored and subsequently reloaded. High register demand may impact performance even in the absence of spillage because it may prevent other optimizations such as pipelining.

To provide accurate and concrete results we show that the performance of a program that makes use of idealized ET code is directly related to the number of registers available; this is compared to code without ETs. Data was collected using the KAI C++ compiler[4] on an SGI Origin 2000 system using the hardware performance counters of the MIPS R10000 microprocessor.

Test Code. Two versions of the test code implement a simple 3-point stencil using multidimensional arrays with subscript computation only along the innermost dimension. The core of the computation is embodied by a loop that traverses the array elements in memory order.

The main differences between the two codes is how the arrays are accessed. In C++ the array on the right-hand side is reused, so resource requirements are constant in the number of operands. However, a code using ETs will transform the statement into a loop that uses a different array pointer for each operand. Further, in an ET code each array carries an integer together with the pointer information on how to compute the proper offset (a more clear idea of the transformation may be had by inspecting the intermediate C code generated by a compiler). We compared the performance of the two codes while independently increasing the number of operands and the dimensionality of the arrays.

Register Spillage Measurements and Results. We capture the impact of register spillage by analyzing the number of loads and stores performed from assembly code and using hardware performance counters. This measure also shows when register demand inhibits other optimizations.

Figure 1 shows that the execution time measured in cycles is significantly different for the two codes. As the number of operands increases the C++ performance stays constant while the ET performance decreases. Secondarily, increasing the dimensionality of the arrays has no effect on the C++ performance but degrades ET performance. In the multidimensional case we see a 60% degradation in ET performance for six dimensions.

The impact of the number of operands increases rapidly starting at 21–26. The MIPS R10000 has 27 general-purpose integer registers, and spilling of integer registers is directly related to the number of operands (there is no spillage of the floating-point registers because the compiler schedules floating-point operations to be performed pairwise). In the ET code two registers are needed for each array operand—one pointer and one for subscript computation. This is detailed in Table 1. For up to four operands ET performance is close to that of C++.

---

[4] The SGI C++ compiler was used to verify the results and conclusions.

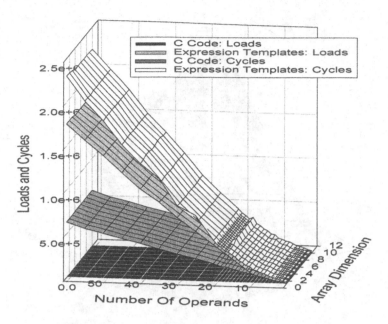

Fig. 1. Register Spillage with Expression Templates

Between five and thirteen operands loop unrolling is inhibited because of register demand. Between 14 and 26 pipelining is inhibited for the same reason. Beyond that register spillage exacts a further penalty.

Table 1. Impact of Register Demand on Pipelining and Unrolling

Number of operands	Fixed-point registers used	Floating-point registers used	Software pipelining	Iterations unrolled
1	15	1	ON	4
2	12	2	ON	2
3	16	4	ON	2
4	12	3	ON	0
5	14	3	ON	0
6	16	4	ON	0
7	18	3	ON	0
8	20	4	ON	0
9	22	4	ON	0
10	24	4	ON	0
11	24	4	ON	0
12	27	5	ON	0
13	27	3	ON	0
14			OFF	0

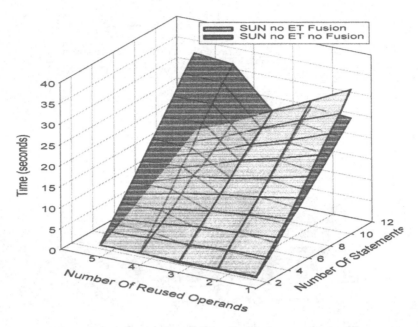

Fig. 2. Loop Fusion

Figure 1 shows the effect of a large number of operands such as appears in 3D codes using geometry where stencils involve connecting cross-derivative terms and variable coefficents. A 3D code could easily use a 27-point stencil and require 27 coefficients, or 54 operands. The C code shows no register spillage, with performance depending only on the number of operands; performance degrades by about 4–5 for 54 operands.

Additional data not presented here shows that the dimensionality of the arrays increases the number of cycles. The difference is a result of the KAI C++ compiler lifting loop-invariant code from an innermost loop to only the next outer loop. The performance degradation is about 60% for 6D arrays, but 6D and higher arrays are important for 3D applications that handle geometry (for example in OVERTURE).

3.2 Loop Fusion as an Optimization Mechanism

Figure 2 shows the effects of loop fusion over a range of 1–12 array statements, parameterized by the number of operands reused within the statements. Where reuse is high loop fusion doubles performance. In C the array statements were written as explicit loops to remove any effects of the array class overhead. We have not shown results in Fig. 2 for 13 and 14 statements for scaling reasons; at this point register spillage causes an order-of-magnitude increase in execution time.

Separate tests of the potential for loop fusion within AMR using red-black relaxation as implemented within an array class have shown factors of 4–6 in improved performance. Similar tests on interpolation operators implemented in the array class yields performance increases of 3–4. From those tests we expect that loop fusion can yield even greater performance gains than presented here.

Similar data for ETs is also available but not presented here. The poor results can be inferred from previous graphs on register spillage with ETs. Loop fusion amplifies this problem by adding even more operands in the inner loop.

Loop fusion is an important optimization that cannot be assured by any combination of C++ compiler or ET technology. This is because the ET implementation must introduce run-time checking for loop dependence so that parallelism (and general array semantics) can be supported. This introduces conditionals between statements (along with hundreds of lines of other ET code, parallel message passing, and so forth), which inhibits compiler-driven loop fusion. This is part of the justification for research on preprocessor mechanisms for loop fusion.

3.3 Stencil and Non-Stencil Applications

A++ implements both binary operator overloading and ETs, so it is a simple setting in which to evaluate the performance of the two mechanisms using the identical application codes. Figure 3 shows the performance of A++ with and without ETs and compares it to C for a 1D shock tube application. While this program does not characterizes all codes (it has no stencil operations, for example), numerical codes are composed of more than stencils. The equivalent performance of the two mechanisms is an unexpected result warranting further investigation.

The results shown in Fig. 4 show that performance is not uniform across all numerical codes. For stencil-like operations the performance of ETs is better than for binary operators, and in fact comes close to that of C code. For stencils additional improvements are possible with blocking schemes, but using them would obscure comparisions with the unblocked C code. The existing implementation is as efficient as the other results for ET implementations, yet for a non-stencil code its performance is no better than the binary operator implementation. We suspect that the computational costs are similar because the number of loads and cache misses are similar and there is little operand reuse on the right-hand sides of the statements.

4 Conclusions

The more traditional execution model of overloaded binary operators has performance problems on cache-based architectures with stencil-type expressions; this is less of an issue on other (e.g. vector) architectures. However, it appears to make little difference if stencils are not an important part of the computation. The ET implementations have extremely poor compile times (factors of 1000 or more), but yield a factor of only 3–4 improvement in operations involving stencils. Using the SGI C++ compiler, where the compile times are only a factor of

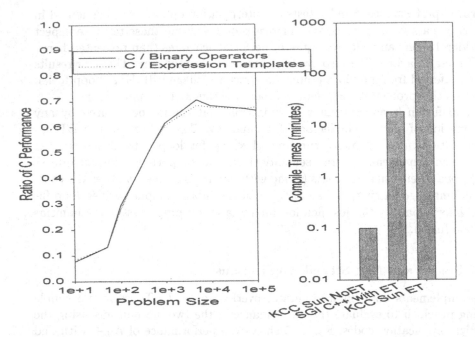

Fig. 3. 1-D Shock Tube Application with Compile Times

20–40 from that of the non-ET code, the performance is far worse than that for binary overloaded operators, for example 1/20 of C performance for 3-D arrays.

ETs also show poor performance for large stencils that appear in OVERTURE applications with complex geometry, though this would not be so much the case in single square grids with no geometry where most of the ET experience is based. In contrast, the binary overloaded operators generate no similar register spillage.

Where the use of ETs significantly improves the performance of array expressions it can still only be compiled by a few C++ compilers (the KAI C++ compiler is the only one which achieves high performance). However, the performance of ETs is sensitive to the quality of the underlying C compiler, something that programmers cannot control.

Finally, we believe that attempts to link the future of object-oriented scientific computing to ET technology are flawed. With some effort better performance can be obtained with a special-purpose optimizing preprocessor. Tools such as SAGE++[3] greatly simplify the development of such preprocessors. More sophisticated transformations are possible with a preprocessor because it can perform semantics-based program analysis impossible with an ET approach. We suspect that preprocessing is a superior and ultimately more powerful alternative to addressing these problems, but enhancement and refinement of the ET mechanism (possibly in conjuction with additional annotation by the user) may also be possible.

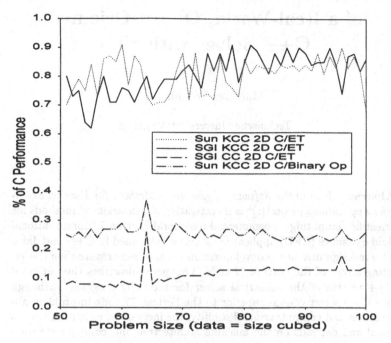

Fig. 4. 2-D Stencil

References

1. David Brown, Geoff Chesshire, William Henshaw, and Dan Quinlan. Overture: An object-oriented software system for solving partial differential equations in serial and parallel environments. In *Proceedings of the SIAM Parallel Conference*, Minneapolis, MN, March 1997.
2. Kei Davis and Dan Quinlan. Rose: An optimizing preprocessor for the object-oriented overture framework. http://www.c3.lanl.gov/~dquinlan/ROSE.html.
3. B. Francois et. al. Sage++: An object-oriented toolkit and class library for building fortran and c++ restructuring tools. In *Proceedings of the Second Annual Object-Oriented Numerics Conference*, 1994.
4. M. Lemke and Dan Quinlan. P++, a c++ virtual shared grids based programming environment for architecture-independent development of structured grid applications. In *CONPAR/VAPP V*, LNCS. Springer-Verlag, September 1992.
5. Dan Quinlan and Rebecca Parsons. Run-time recognition of task parallelism within the p++ parallel array class library. In *Proceedings of the Conference on Parallel Scalable Libraries*, 1993.
6. Dan Quinlan and Rebecca Parsons. A++/p++ array classes for architecture independent finite difference computations. In *Proceedings of the Second Annual Object-Oriented Numerics Conference*, April 1994.
7. Todd Veldhuizen. Expression templates. In S.B. Lippmann, editor, *C++ Gems*. Prentice-Hall, 1996.
8. O.V. Wilson and P. Lu. *Parallel Programming Using C++*, chapter 14, pages 547–587. MIT Press, 1997.

Design and Performance Improvement of a Real-World, Object-Oriented C++ Solver with STL

Matthias Weidmann

Technische Universität München

Abstract. Part of the *Software Engineering Methods for Parallel Scientific Applications* project[1][1] is investigating object-oriented methods for scientific computing. A commercial real-world Fortran 77 computational fluid dynamics (CFD) application is being redesigned in C++ and Java. This redesign uses newly developed numerical objects rather than the existing arrays of the Fortran 77 code. This paper describes the improved C++ version of the numerical solver for the CFD program. Although this C++ solver code is superior to the Fortran 77 code in problem abstraction and computational flexibility, the increase in execution time is small and depends on the machine architecture and compiler software used.

1 Introduction

This is a continuation of the work presented in [2], where the object-oriented redesign of the algebraic multigrid (AMG) solver [3] of the CFD software package *TASCflow for CAD* (*CFX-TfC*)[2] and a first AMG implementation in C++ were discussed. The results there show a superior quality for the C++ AMG code compared to the Fortran 77 AMG code, but also show a disastrous increase rate in execution time that grows linearly with the complexity of the test cases. This increase was implementation-dependent and not related to the use of object-oriented software techniques.

In this paper, an improved C++ AMG implementation is described. Design improvement was achieved by using components of the *Standard Template Library* (STL) [4]. Performance improvement was achieved by eliminating work the previous C++ AMG code performed in addition to the Fortran 77 AMG code: unnecessary dynamic memory allocation was eliminated, and the previously used sparse matrix data structure (proved inefficient for the AMG algorithm) was replaced.

---

[1] Funded by the German Federal Department of Education, Science, Research and Technology.
[2] From AEA Technology GmbH, info@ascg.de.

2 The AMG Solver

CFX-TfC's AMG solver is an implicit coupled iterative solution approach. The equation systems to be solved are derived from an unstructured grid, finite-volume discretization method of the 3D Navier-Stokes and scalar transport equations. Therefore, the AMG solver must handle two different equation systems with differing structure. In the coupled equation system, matrix and vector elements themselves are matrices and vectors, whereas the scalar equation system is a normal matrix and vector system with one-dimensional matrix and vector elements. This means that the requirements on the numerical data structures are more sophisticated for the coupled equation case than for the scalar case. Differently dimensioned matrix and vector element data structures must be available and handled accordingly by the AMG method implementation at run time. The Fortran 77 AMG solver handles both equation system cases with one code implementation by intelligent array indexing. Similarly, the C++ AMG solver has been designed to treat both equation systems with one code implementation.

3 Class Design Improvement

The class design improvements proposed in [2] were carried out. Figure 1 shows the new class design for the general equation system solver. Further, the basic numerical objects *Sparse Matrix*, *Main Diagonal* and *Vector* are now based on the STL components *map* and *vector*. The motivation to use STL for class design came from the need to create a new sparse matrix data structure for efficiency reasons. The required special properties of the sparse matrix data structure could be met easily with the STL components *map* and *vector*. The special STL sparse matrix implementation is described in Sec. 4.2. In redesigning the sparse matrix data structure with STL it was noticed that the *Main Diagonal* and *Vector* objects could profit by *vector* as well. This led to a reduction of over 30% in lines of C++ code.

The leftmost objects in Fig. 1, *Matrix Element* and *Vector Element*, implement the general equation system case. This allows one code formulation for both the fully coupled equation system and the scalar equation system. Unfortunately, this elegant class design suffers from run-time inefficiency owing to object management overhead, at least for the scalar equation system case, because the objects *Matrix Element* and *Vector Element* are just floating-point values. For the scalar equations, Sec. 4.3 shows how this class design is optimized easily using the STL.

The high-level AMG objects are nearly unchanged. The management of the sparse matrix in the *Algebraic Grid* object became easier because of the new sparse matrix structure.

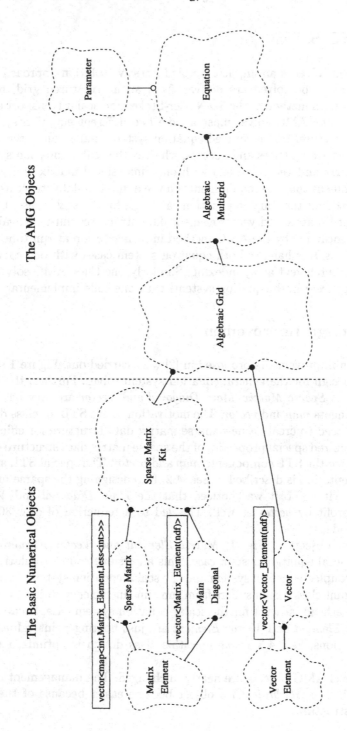

Fig. 1. The STL-based class diagram of the C++ AMG solver.

4 Performance Improvements

In this section we describe performance improvements owing to the elimination of unnecessary dynamic storage allocation, to an efficient sparse matrix data structure, and to optimizing the class design for the scalar equation.

4.1 Eliminating Dynamic Allocation in the Equation System

The previous C++ AMG code allows a flexible way of solving equation systems as the number of coupled equations was not fixed until run time. Accordingly, the coupled and scalar equation systems are handled at run time. In the objects *Matrix Element* and *Vector Element* (see Fig. 1), dynamic memory allocation is done for each matrix or vector element, respectively, with the C++ system call *new*. Dynamically allocating individual *float* values at the lowest abstraction layer yields a memory-efficient implementation for arbitrary coupled equations, but results in a high execution time.

However, this flexible solution structure is not necessary. In practice it is unreasonable to start from the assumption that the type of the equation system is not known. The degree of coupling in the equations to be solved is known in advance and can be used for allocation of the matrix and vector element data structures at compile time, leading to a fixed-size matrix and vector element allocation. For the coupled case, this is memory efficient. For the scalar case, this is not memory efficient, as more memory is used for each matrix and vector element than necessary. However, this is only true if the AMG method has one program code which solves the coupled equations as well as the scalar equations.

If two AMG codes for the coupled and the scalar equations were implemented (differing in the size of the matrix and vector elements), the scalar AMG code implementation can also be memory efficient. But the overall maximum memory requirement of *CFX-TfC* would not decrease if this were done. So, no memory-efficient AMG code is really needed, and dynamic memory allocation can be eliminated.

4.2 Providing an Efficient Sparse Matrix Data Structure

A new object-oriented algebraic multigrid sparse matrix data structure based on STL has been implemented in the improved AMG code to meet two requirements: efficient reading of the matrix elements, and efficient dynamic growth for the generation of the coarse grid matrices at run time. Efficiently building the coarse grid sparse matrices is critical, because their structure is not known in advance. The dimension of a coarse grid sparse matrix is computed at run-time, but the exact number of non-zero elements is only known after the coarse grid matrix has been built. This is because the coarse grid matrix elements are formed algebraically in a random number and order and, furthermore, identical coarse grid matrix elements can be formed and inserted several times in the coarse grid sparse matrix. This requires an efficient existence-check of matrix elements and an efficient change mechanism for the matrix structure. The coarse grid sparse matrix data structure therefore must allow for direct in-place insertion.

In the improved AMG code, the sparse matrix data structure is implemented as a STL *vector* of dynamic STL *maps* (see the *Sparse Matrix* object in Fig. 1):

```
typedef map<int,Matrix_Element,less<int>>::iterator mat_it;

class Sparse_Matrix_Kit
{
private:
        vector<map<int,Matrix_Element,less<int>>> Sparse_Matrix;
        vector<Matrix_Element> Main_Diagonal;
public:
        . . .
};
```

This data structure provides an efficient, direct, in-place insertion and therefore supports generation of the coarse grid matrices with constant effort. This was not the case in the previous C++ AMG implementation where the coarse grid matrices are generated inefficiently, in $O(n^2)$ time. Now the execution time for a test problem is independent from the problem size and remains constant. This change is the principal contributor to performance improvement.

4.3 Optimizing the Class Design for the Scalar Equation

For an efficient implementation of the scalar case, the objects *Matrix Element* and *Vector Element* must be replaced by simple floating point values. This is achieved easily within the current design by replacing the types *Matrix Element* and *Vector Element* in the *vector* and *map* templates with the *float* data type (the modified classes are shown in Fig. 2). The STL based sparse matrix definition is reduced to:

```
typedef map<int,float,less<int>>::iterator mat_it;

class Sparse_Matrix_Kit
{
private:
        vector<map<int,float,less<int>>> Sparse_Matrix;
        vector<float> Main_Diagonal;
public:
        . . .
};
```

5 Results

In this section we discuss how the changes described in the last section impacted code characteristics and execution performance.

The Basic Numerical Objects

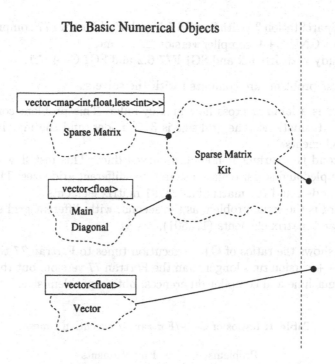

Fig. 2. The optimized classes for the scalar equation.

5.1 Design

The new AMG class design based on STL has led to an reduction of over 30% in programmed C++ code lines, compared to the previous C++ AMG class design. Now the C++ code has 6547 lines, compared to 3966 lines of Fortran 77 code.

The performance improvements in the C++ AMG code required changes in the basic numerical objects. This did not affect the pseudo-code-like implementation of the actual AMG algorithm, which is formulated in the high-level object *Equation* (see Fig. 1). The pseudo-code-like C++ AMG code is robust regarding implementation changes of the underlying numerical objects due to their encapsulated formulation.

5.2 Performance

For the test runs, the implementation of the general class design is used for the coupled equation system and the implementation of the specialized class design (Sec. 4.3) is used for the scalar equations.

The test environments are (all compilers are driven with -O3 optimization level):

A. a SUN Sparcstation 10 with SunOS 4.1.1 and Fortran 77 SUN SC compiler version 2.0.1 and GNU C++ compiler version 2.7.1;

B. a SUN SparcStation 20 with SunOS 5.5 with PGI Fortran 77 compiler version
   1.4-3 and GNU C++ compiler version 2.7.2; and
C. a SGI Indy with Irix 6.2 and SGI F77 6.2 and SGI C++ 7.1.

Three test problems are computed with the solvers.

1. The first is a laminar cross flow in a cylinder. This problem consists only
   of coupled equations. The grid size is 3150 nodes, and the matrix has 8865
   matrix elements.
2. The second is a turbulent flow in a curved duct. This test flow consists of
   both coupled and scalar equations. and two different grid sizes: The grid size
   is 3627 nodes, and the matrix has 23791 matrix elements.
3. The third is the same problem as the second, with different grid size (25925
   nodes) and matrix elements (175501).

Table 1 shows the ratios of C++ execution times to Fortran 77 times. In all
cases the C++ version runs longer than the Fortran 77 version, but the increases
depend on machine and compiler differences, not on problem size.

Table 1. Ratios of C++/Fortran 77 execution times.

Problems	Environments		
	A	B	C
1 (coupled eqs)	2.5	6.8	3.4
2 (coupled eqs)	—	6.5	3.4
2 (scalar eqs)	—	1.5	4.2
3 (coupled eqs)	—	6.5	3.4
3 (scalar eqs)	—	1.5	4.2

6 Conclusions and Future Work

The AMG solver algorithm profits from the object-oriented redesign in C++ in
terms of its pseudo-code-like implementation and the flexible algorithm management at run-time due to dynamic object creation. The C++ code is superior to
the Fortran 77 code in these ways:

- The readability of the numerical algorithm is enhanced. It is robust regarding code changes due to object encapsulation and hence maintainability is
  enhanced.
- The C++ code can adjust itself to the memory requirements at run time
  during the generation of the coarse grid matrices. The Fortran 77 AMG code
  must be recompiled if the assumed workspace memory size turns out to be
  too small.

- The greater execution times of the C++ AMG code include the overhead of the dynamic memory allocation during the generation of the AMG coarse grid matrices at run-time. This expensive feature leads to an increased AMG code stability, but without it, the C++ code would achieve better performance. In addition, Fortran 77 compilers are typically more mature, and have an easier job in optimization for numerical applications. C++ compilers are still being improved, and better optimizations are in progress.

The computation of the coupled equations could be optimized if improvements to the general class design for scalar equations were also applied: The elimination of the objects *Matrix Element* and *Vector Element* would reduce object management overhead at run-time.

The AMG objects (on the right hand side in Fig. 1) could also profit from a STL based design improvement. The *Algebraic Multigrid* object represents the entire multigrid hierarchy as a list of *Algebraic Grid* objects. It could be replaced with the STL template *list* in the *Equation* object instead.

A Java AMG solver based on this class design is under development.

References

1. Peter Luksch, Ursula Maier, Sabine Rathmeyer, Matthias Weidmann, and Friedemann Unger: *Sempa: Software Engineering Methods for Parallel Scientific Computing*, IEEE Concurrency, July-September 1997.
2. M. Weidmann: *Object-Oriented Redesign of a Real-World Fortran 77 Solver*, chapter in "Modern Software Tools for Scientific Computing", edited by Erlend Arge, Are Magnus Bruaset and Hans Petter Langtangen, Birkhäuser 1997, http://www.birkhauser.com/cgi-win/isbn/0-8176-3974-8.
3. Michael. J. Raw: *A Coupled Algebraic Multigrid Method for the 3D Navier-Stokes Equations*, in "Fast Solvers for Flow Problems", Proceedings of the 10th GAMM-Seminar, Notes on Numerical Fluid Mechanics Vol. 49, Vieweg-Verlag, Braunschweig, Wiesbaden, 1995.
4. Musser David R., Saini Atul: *STL Tutorial and Reference Guide: C++ Programming with the Standard Template Library*, Addison-Wesley, 1996, http://www.aw.com/cp/musser-saini.html

Evaluating High Level Parallel Programming Support for Irregular Applications in ICC++*

Andrew A. Chien, Julian Dolby, Bishwaroop Ganguly, Vijay Karamcheti, and Xingbin Zhang

University of Illinois at Urbana-Champaign

1 Overview

Object-oriented techniques have been proffered as aids for managing complexity, enhancing reuse, and improving readability of parallel applications, and numerous variants have been explored [14]. These techniques appear especially promising for irregular applications, where modularity and encapsulation help manage complex data and parallelism structures. In this paper, we evaluate a high level concurrent object system's programmability for achieving high performance on irregular applications. Prior research exists for both implementing irregular parallel applications efficiently and concurrent object-oriented programming, but we know of no studies that address both together.

Of the research on programming irregular parallel applications, none provides a single programming interface while supporting all three dimensions of irregularity: data, control, and concurrency. Research efforts have largely focused on building libraries which provide run-time support for some dimensions of irregularity [6, 9, 13]. Most concurrent object-oriented programming systems have ignored support for fine-grained parallelism [2, 5, 8], which forces programmers to map irregular concurrency into grains large enough to achieve efficiency.

Our study is done in the context of the Illinois Concert system, a high-performance compiler and run-time system for parallel computers, which has been the vehicle for research on compiler optimization and run-time techniques over the past five years [3]. Using the Concert system, we evaluate programming effort for a varied suite of irregular parallel applications.

Our results show that a high-level, concurrent, object-oriented programming model and a sophisticated implementation can eliminate many concerns, easing the programming of irregular parallel applications. First, programmers are freed from explicit namespace management, computation granularity (procedure and data), and low-level memory race management. Second, while in most cases, programmers still needed to express application-specific data locality and load balance, the orthogonal framework for functionality and data placement eased

* This research was supported in part by DARPA#E313 through AFOSR Contract F30602-96-1-0286, ONR grants N00014-92-J-1961, N00014-93-1-1086, NASA grant NAG 1-613, and T3D resources at the Jet Propulsion Laboratory. Andrew Chien is supported in part by NSF YI Award CCR-94-57809. Vijay Karamcheti was supported in part by an IBM Computer Sciences Fellowship.

this effort by allowing radical changes in data layout with modest code changes. Finally, over the seven applications only modest code changes (fewer than 5% of the lines) were required, mostly to express the careful data locality and load balance needed to achieve top performance.

2 Background

In evaluating a programming approach, our goal is to identify the concerns which must be addressed by a programmer to achieve good performance and how those concerns are managed. Two high-level concerns are of most interest: *concurrency and synchronization specification*, and *data locality and load balance*. The former is required to express the parallelism structure, while the latter ensures efficient execution. We describe interfaces in the Concert system which aid in the management of these two concerns.[1]

Basic Programming Interface   Our programming interface is an object-oriented programming model augmented with simple extensions for concurrency. The model is based on C++ (ICC++ [14]), providing a single namespace, scalars and objects, and single inheritance. Simple extensions for concurrency include annotating standard blocks (compound statements) and loops with the conc keyword. A conc block defines a *partial* order among its statements, allowing non-binding concurrency while preserving local data dependences. A simple semantics for object concurrency provides synchronization: method invocations on an object execute as if they had exclusive access.

Performance Optimization Interfaces   Data placement and colocation is managed through *collections*, distributed arrays of objects for which the user can specify standard layouts (block, cyclic) or full custom distributions. Between individual objects, colocation control can be expressed at object allocation time.

Three additional interfaces support control over data consistency, thread placement, and task ordering. Whereas the data placement interfaces were found useful in all applications, these additional interfaces are only required by a few of the applications. The data-consistency interface allows the Concert system to use relaxed consistency models for higher performance. As shown in the following code fragment, the with_local annotation specifies a set of objects that need to be localized and provides information about read-only access semantics:

```
with_local( obj1, READ_ONLY ){ ... = func( obj1, ... );}
```

The thread placement interface provides support for controlling data locality and load balance from the perspective of threads, and is specified by annotating thread creation with the !target keyword and an integer argument. In this fragment, the thread corresponding to thread_function is executed on a random processor.

```
( thread_function( arg1, arg2, ... ) !target rand() );
```

---

[1] More information about Concert is available at http://www-csag.cs.uiuc.edu/

Task ordering is useful in applications based on search which prioritize promising tasks for higher efficiency. Task ordering is specified by annotating the thread creation site using the !priority keyword, which takes the task priority as an argument. Additionally, the interface permits the programmer to specify a custom scheduler for thread execution. In the following fragment, the thread corresponding to thread_function executes with priority arg1 and is managed by a global priority scheduler:

```
(thread_function(arg1, ...) !priority arg1 !scheduler PRIORITY);
```

Implementation Technology   The Concert system contains a wide range of aggressive optimizations, and has been used to show high performance in absolute terms on a wide range of applications [3]. It automatically addresses several concerns which programmers must manage in lower level programming models: procedure granularity and virtual function call overhead; compound object structuring; thread granularity; distributed name management; initial data placement, data consistency and sharing.

3   Application Case Studies

We studied the benefits of high-level programming features and support using a suite of seven applications (see Table 1). Each problem is challenging to express in parallel, exhibiting irregularity in one or more of data structure, parallel control structure, and concurrency.

In this paper we discuss two of them: Gröbner and and Radiosity.[2] For each application, we describe programmer effort associated with concurrency specification and ensuring data locality and load balance.

3.1   Gröbner

This application computes the Gröbner basis of a set of polynomials. Our algorithm is derived from a parallel implementation by Chakrabarti [1].

Table 1. The irregular applications suite

Applications	Description
Polyover [14]	Overlay of two polygon maps (computer graphics)
Barnes-hut [15]	Barnes-Hut hierarchical N-body method (computational cosmology)
FMM [15]	Fast multipole method (computational cosmology)
SAMR	Structured adaptive mesh refinement (computational fluid dynamics)
Gröbner [1]	Gröbner basis (symbolic algebra)
Phylogeny [7]	Evolutionary history of species (molecular biology)
Radiosity [15]	Hierarchical radiosity method (computer graphics)

[2] See a longer version of this paper at http://www-csag.cs.uiuc.edu/ for the rest.

Concurrency and Synchronization Specification   Parallelism in the application arises from parallel evaluation of polynomial pairs, and is naturally specified using a conc loop around the pair-generation code in the sequential program. Pair-evaluation tasks that produce irreducible terms invoke a method on the basis object to augment it. Global namespace support in Concert ensures that potentially remote objects (either the polynomial or the basis) can be accessed with code identical to the sequential methods. The only synchronization speci-fication required is that the basis object be kept consistent against concurrent access, ensured in Concert due to the object-level concurrency control semantics. The natural expression of concurrency and synchronization specifications only require changes to 5 lines in the program (out of 1919 ICC++ lines).

Data Locality and Load Balance   Pair evaluation work is data-dependent and varies with evaluation order. Consequently, careful management of data locality and load balance is required for efficiency.

Data locality ensures that the pair tasks have efficient access to the basis and its polynomials. Since load balancing requires pairs to be evaluated on arbitrary processors, both polynomial and basis objects are cached on demand, utilizing additional information about relaxed data consistency semantics. Programmer input is required for this information which conveys the fact that polynomials are accessed in a read-only fashion, and that the computation can proceed with an inconsistent view of the basis. The latter permits overlap of pair evaluations with basis augmentations, and is expressed as in the following code segment. Over the entire program, the programmer needs to specify access information about 2 kinds of views which requires changes to 4 lines in the program.

```
processPair( pOne, pTwo ) {
  spol = spol( pOne, pTwo );
  basisp = &(Basis);   /* can use an inconsistent Basis copy */
  with_local}( basisp, READ_ASYNC )
    { ... = basisp->setReduce( spol ); }
}
```

Load balance is needed for pair-evaluation tasks which also require priori-tized execution to avoid redundant work. Our implementation attaches the pair evaluation task to a priority-based, work-stealing scheduler (provided as part of the Concert system) as shown in the code fragment below. The programmer must identify the task granularity (the processPair function), supply an integer priority using the !priority keyword, and specify a scheduler for enqueuing the task (the GLOBAL_PRIORITY scheduler). Similar custom load-balancing annota-tions were required to two lines in the program.

```
conc for ( pOld = ... ) {
  if ( (pOld != pNew) && test(pOld, pNew) )
    (processPair(pOld,pNew) !priority calculate_priority(pOld,pNew)
        !scheduler} GLOBAL_PRIORITY );
}
```

3.2 Radiosity

The radiosity method computes the global illumination in a scene. Our parallel algorithm is derived from the radiosity application in the SPLASH-2 suite [15].

Concurrency and Synchronization Specification   There are three levels of parallelism in each iteration: across all input patches, across children of a subdivided patch, and across neighbor patches in the interaction list. These are all naturally expressed using conc annotations. Concert's global namespace enables access to remote objects with code identical to sequential methods. Two synchronization specifications ensure correct execution, expressed using the object-level concurrency control semantics. The first specifies that a child patch radiosity calculation must be nested within that of its parent. The second specifies exclusive access to the patch object. Over the entire program, these specifications are only required in seven places (out of 3144 lines).

Data Locality and Load Balance   This application has data-dependent irregularity at each level of parallelism as well as across iterations. So careful management of data locality and load balance is required. Data locality is required for radiosity and visibility tasks. The former accesses the target patch and, for each interaction, the interaction object and the neighbor patch. The visibility task accesses a pair of patches, and traverses the BSP tree in a data-dependent fashion. Locality of interaction objects and the BSP tree is ensured by data alignment: the former with the source patch and the latter by explicit replication. In contrast, patch locality requires demand-driven caching, which is made efficient by using relaxed consistency information. As seen in this fragment, the programmer uses the with_local annotation to specify the read-only access semantics (a view) of both polygonal input patches. The programmer must specify similar access information about three kinds of views. Over the entire program, ensuring data locality increases the source code length by 33 lines (as compared to the original 3144 lines): of these the bulk (25) replicates the BSP structure, and the rest describe alignment and data consistency specifications.

```
computeVisibility( spatch, dpatch, cur ) {
    with_local( spatch, READ_ONLY, dpatch, READ_ONLY )
        { vis = vis_func( spatch->pos, dpatch->pos, ... ); }
    cur->vis = vis;
}
```

Load balance is required to balance the radiosity and visibility tasks across the machine. Allocating dynamically created patches on random nodes at creation time (the default policy in Concert), and executing radiosity tasks local to them ensures that the latter is load balanced. However, load balancing visibility tasks requires a sender-initiated scheme. As the following code fragment shows, the programmer had to identify the task which must be balanced (computeVisibility) and its target destination (a random node), both of which are specified in Concert using the !target keyword. Similar annotations were required at three places in the program.

```
conc for (cur=start_interaction_list; cur; cur = cur->next ) {
    if ( cur->vis == VISIBILITY_UNDEF )
        ( computeVisibility( this, cur->dest, cur ) !target rand() );
}
```

3.3 Summary

All of the applications use globally-shared, pointer-based data structures, and most have irregular concurrency structures, which are expressed naturally using the global namespace, object consistency model and conc annotation of ICC++.

The application suite falls into two main classes. Polyover, Barnes-Hut, FMM, and SAMR exhibit irregular concurrency which requires either semi-static or random load balancing schemes. Programmer input is only required in the form of a small amount of code for load balancing and data placement, typically using collection maps. As a result, fewer than 5% of the lines typically require changes: 81 out of 1695 for Polyover, 107 out of 2631 lines for Barnes-Hut, 163 out of 3951 lines for FMM and 163 out of 6303 lines for SAMR.

On the other hand, Radiosity, Gröbner, and Phylogeny additionally require fully dynamic load balancing, application-specific scheduling and relaxed consistency models. Thus, additional effort is required from the programmer who must have intimate knowledge of the application to specify load balancing policies, thread priorities, and data-access semantics. However, Concert provides high-level annotations to allow these kinds of information to be incorporated into the computation with very little code change: 11 out of 1919 lines for Gröbner, 16 out of 2305 lines for Phylogeny, and 43 out of 3144 lines for Radiosity.

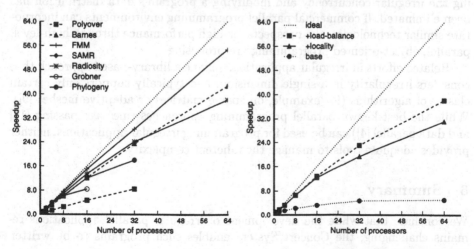

Fig. 1. Overall speedup of application suite (left), and Incremental performance benefits from program Transformations for the Radiosity application (right)

4   Parallel Performance

All programs[3] achieve good sequential performance (within a factor of two of corresponding C programs), benefiting from implementation techniques that eliminate programmer management of concerns such as procedure and thread granularity, and object structuring, otherwise required for performance. Figure 1(left) shows speedups with respect to the non-overhead portion of the single-node execution time. They compare favorably with the best reported elsewhere for hand-optimized codes. Polyover's speedup levels off because of insufficient work in the parallel algorithm, but is comparable to other reported numbers [14]. Both the Barnes speedup of 42 and FMM speedup of 54 on 64 nodes are competitive with shared-memory [15] and hand-optimized versions [12]. The Radiosity speedup of 23 on 32 T3D processors compares well with a reported speedup of 26 on 32 processors of the DASH machine [10]. Speedups for both Gröbner and Phylogeny compare favorably with implementations that explicitly manage data locality [1, 7].

These speedups were enabled by transformations that enhance data locality and load balance. To show their performance advantages, Fig. 1(right) shows the speedup for three Radiosity versions: base, locality, and load-balance. Both locality and load-balance versions are require changes to less than 2.5% of the source code, yet improve speedup significantly.

5   Discussion and Related Work

To our knowledge this the only study which looks at a collection of irregular applications on a general-purpose parallel programming system to evaluate its support for high-level expression and performance optimization. We find the outcomes encouraging. In the Concert system, much of the difficulty of managing the irregular concurrency and modifying a program's data distribution has been eliminated. If commercial parallel programming environments can incorporate similar technologies, the prospects for high performance through high-level, parallel, object-oriented programming are promising.

Related efforts in irregular applications pursue a library-based approach that considers irregularity in a single dimension, and typically support only certain classes of algorithms (for example, on sparse matrices or adaptive meshes [9]). While the best-known parallel programming approaches (message passing [11] and data parallel [4]) can be used for programming irregular applications, neither provides adequate tools to manage the inherent complexity.

6   Summary

We conclude that while the development of irregular parallel applications remains challenging, the Concert System enables such programs to be written

---

[3] Results are based on the Cray T3D implementation of Concert, except for the SAMR code, whose performance results are on the SGI Origin 2000.

with a high-level approach. Although the complexity of data locality and load balance is significant (often comparable to the algorithmic complexity), the high-level program expression enables their orthogonal expression, improving program flexibility. Finally, the optimized parallel versions of the irregular applications had the same basic program structure as their sequential counterparts. Concurrency was added with small program perturbation, and data locality and load management was achieved with minor program changes.

References

1. S. Chakrabarti and K. Yelick. Implementing an irregular application on a distributed memory multiprocessor. In *Proc. 4th ACM/SIGPLAN Symp. on Principles and Practices of Parallel Programming*, pp. 169–179, May 1993.
2. K. Mani Chandy and C. Kesselman. Compositional C++: Compositional parallel programming. In *Proc. 5th Workshop on Compilers and Languages for Parallel Computing*, New Haven, Connecticut, 1992. YALEU/DCS/RR-915, Springer-Verlag Lecture Notes in Computer Science, 1993.
3. A. Chien, J. Dolby, B. Ganguly, V. Karamcheti, and X. Zhang. Supporting high level programming with high performance: The Illinois Concert system. In *Proc. 2nd Int. Workshop on High-level Parallel Programming Models and Supportive Environments*, April 1997.
4. High Performance Fortran Forum. High performance Fortran language specification version 1.0. Technical Report CRPC-TR92225, Rice University, Jan. 1993.
5. A. Grimshaw. Easy-to-use object-oriented parallel processing with Mentat. *IEEE Computer*, 5(26):39–51, May 1993.
6. J.H. Saltz, *et al.* A manual for the CHAOS runtime library. Technical Report CS-TK-3437, Computer Science Dept., U. Maryland, 1995.
7. J. Jones. Parallelizing the phylogeny problem. Master's thesis, Computer Science Division, University of California, Berkeley, Dec. 1994.
8. J. Lee and D. Gannon. Object oriented parallel programming. In *Proc. ACM/IEEE Conf. on Supercomputing*. IEEE Computer Society Press, 1991.
9. R. Parsons and D. Quinlan. A++/P++ array classes for architecture independent finite difference computations. In *Proc. 2nd Annual Object-Oriented Numerics Conf.*, Sunriver, Oregon, April 1994.
10. J. Singh. *Parallel Hierarchical N-Body Methods and Their Implications For Multiprocessors*. PhD thesis, Stanford University Computer Science Dept., Stanford, CA, February 1993.
11. M. Snir, et. al. *MPI: The Complete Reference*. MIT Press, 1995.
12. M. Warren and J. Salmon. A parallel hashed oct-tree N-body algorithm. In *Proc. Supercomputing Conf.*, pp. 12–21, 1993.
13. C.-P. Wen, S. Chakrabarti, E. Deprit, A. Krishnamurthy, and K. Yelick. Run-time support for portable distributed data structures. In *3rd Workshop on Languages, Compilers, and Run-Time Systems for Scalable Computers*, pp. 111–120, Boston, May 1995. Kluwer Academic Publishers.
14. G. Wilson and P. Lu, eds. *Parallel Programming Using C++*. MIT Press, 1995.
15. S. Woo, M. Ohara, E. Torrie, J. Singh, and A. Gupta. The SPLASH-2 programs: Characterization and methodological considerations. In *Proc. Int. Symp. on Computer Architecture*, pp. 24–36, 1995.

Processing Sparse Vectors
During Compile Time in C++

Jaakko Järvi

Turku Centre for Computer Science, Finland

Abstract. A C++ template library for a special class of sparse vectors is outlined. The sparseness structure of these vectors can be arbitrary but must be known at compile time. In this case it suffices to store only the nonzero elements of the vectors, and no indexing information about the sparseness pattern is required. This information is contained in the type of the vector as a non-type template parameter. It is shown how common vector operators can be overloaded for these vectors. When compiled the operators yield code which performs only the necessary elementary operations between the nonzero elements with no run-time penalty for indexing. All indexing is performed at compile time, resulting in very fast execution speed. The vector classes are best suited for short vectors up to few dozens of elements.

Automatic differentiation of expressions is given as an example application. It is shown how classes for automatically differentiable numbers can be defined with the library. A comparison against other vector representations gave superior results in execution speed of differentiating a few common expressions, and came very close to the calculation speed of symbolically differentiated expressions.

1 Introduction

Templates are a powerful feature of C++. They are usually used to write generic classes and functions, but we can go beyond that. It is possible to make the compiler perform as an interpreter at compile time. For instance, compile-time bounded loops and branching statements can be written using recursive template definitions and template specialisation. These templates are called *template metaprograms* [1]. The use of non-type template parameters is a key factor behind template metaprograms.

In this article template metaprograms are used in the definition of *compile-time sparse vector* (CTSV) classes. The term *compile-time* here means that the sparseness pattern of the vectors (the positions of zero and non-zero elements) is known at compile time. Due to this restriction CTSVs do not serve as general purpose sparse vectors, but they are very efficient in special applications where the above requirement can be satisfied.

The definitions for CTSV template classes are given. It is shown how common operations can be overloaded for CTSV types. Automatic differentiation [2] is presented as an application of CTSVs. The techniques presented take advantage of the new template features present in the current draft standard of C++ [3], but are not yet implemented in all compilers. A more detailed discussion can be found in [4].

2 Class Templates for Compile-Time Sparse Vectors

A vector is called sparse if only a few of its elements are nonzero. In sparse storage schemes only the nonzero elements are stored, along with some auxiliary information to determine the logical positions of the elements in the vector. Hence space is saved and computational speed gained since some elementary operations are performed only on nonzero elements. An arbitrary sparse vector can be written as a set of value-position pairs

$$x = (< x_{i_1}, i_1 >, < x_{i_2}, i_2 >, \ldots, < x_{i_n}, i_n >).$$

With this notation we can write the sum of two sparse vectors $(1, 0, 1, 0, 0)$ and $(0, 0, 2, 0, 2)$ as

$$(< 1, 1 >, < 1, 3 >) + (< 2, 3 >, < 2, 5 >) = (< 1, 1 >, < 3, 3 >, < 2, 5 >).$$

As can be seen, instead of five additions between the elements, only one addition and two value copy operations are needed to compute the result. To perform only the necessary elementary operations, some kind of indexing scheme must be used to find the right operands. This bookkeeping causes extra overhead which we want to minimise. In CTSVs the bookkeeping can be avoided totally, since it is done by the compiler with no run-time penalty. The indexing information is contained in the template parameters of CTSV classes. In other words, each vector with a different sparseness pattern is a type of its own.

Template definitions can become lengthy, so the code in this article is given for floating-point vectors instead of generic vectors. It is straightforward to generalise the class definitions by making the element type a template parameter.

A special representation for a vector element is needed:

```
template <unsigned int N> class Elem {
public:
    Elem<N> (float v) {}
    Elem<N> () {}
};
template<> class Elem<1> {
public:
    float value;
    Elem<1>(float v) : value(v) {}
    Elem<1>() {};
};
```

The class Elem has a template parameter N, and the class definition for any N is a class having no data members, just two constructors. For N=1 a specialisation is provided containing a data member for storing a value. So Elem<0> is an empty class, and Elem<1> is a class containing a single floating-point value. Other values than 0 or 1 for N are not meant to be used. The default constructor does nothing. The constructor taking a floating-point parameter initialises the element with a value. To give a uniform interface, the class Elem<0> also has a constructor taking a floating-point value, though it performs no action.

The class representing a CTSV is a collection of Elem<N> objects: Elem<1> objects in nonzero positions and Elem<0> in zero positions. With this kind of element type definitions, we are able to store empty classes as the zero elements. Note however that an object of an empty class may not be totally empty. It is common for a single unused byte to be allocated.

The sparseness pattern of a vector x can be characterised with a bit sequence b_x, where 1 corresponds to a nonzero element and 0 to a zero element. Since integral types can be manipulated as bit patterns in C++, an unsigned integer template parameter can be used to represent the bit sequence in the generic CTSV class definition. The bit pattern of this template parameter determines the nonzero positions of the CTSV. The class definition is:

```
template <unsigned int N> class CTSV {
public:
  Elem< N&1 > head;
  CTSV< N>>1 > tail;
  CTSV<N>(float v) : head(v), tail(v) {}
  CTSV<N>() {};
  CTSV< N>>1 >& GetTail() { return tail; }
};
template<> class CTSV<0> {
public:
  Elem<0> head;
  CTSV<0>(float v) : head() {};
  CTSV<0>(){};
  CTSV<0>& GetTail() { return *this; }
};
```

The definition is recursive. The head member holds the value of an element. Whether it will be of type Elem<0> (zero element) or Elem<1> (nonzero element) is determined by the least significant bit of the template parameter N. This can be examined by taking the bitwise AND operation with 1. The tail member holds the remaining part of the CTSV. The value of the template parameter for the next step of the recursion is given by right-shifting N one bit. This will eventually result in the template parameter being zero, and the specialisation for N = 0 ends the recursion.

The default constructor is defined to do nothing. In addition, a constructor taking a single float argument is provided for initialising a vector with a given value. It will pass the same argument to the head and tail members. In the case of Elem<1> type head, the value is stored, otherwise nothing is done, since the respective Elem<0> constructor is empty. The recursion is ended with the empty constructor of CTSV<0>. Since CTSV<0> objects have no tail member, GetTail functions are provided. They are needed in the operator definitions to allow uniform access to the vector tail. We may need to get the tail of a CTSV<0> object, and a tail of a tail of a CTSV<0> object, and so on.

The above class definition generates many function calls. Thus it is crucial that all functions be inlined, so empty functions are discarded from the compiled code.

2.1 CTSV Operations

The most common mathematical operations defined for vectors (addition, subtraction, unary negation, multiplication by a scalar, and dot product) can be implemented easily. The definition of addition for CTSVs is given below. Consider two sparse vectors x and y with bit sequences b_x and b_y. Vector $x + y$ has then characteristic bit sequence b_x OR b_y. In C++ this is:

```
template <unsigned int N, unsigned int M>
inline CTSV<N|M> operator+(const CTSV<N>& a, const CTSV<M>& b) {
    CTSV<N|M> c; plus<N,M>::add(a,b,c); return c;
};
```

This template can be instantiated with two CTSVs having arbitrary bit sequences. The resulting type is a CTSV having a characteristic bit sequence formed by bitwise OR. The operator+ serves as an interface to the actual addition operation implemented as a static member function add of a generic class plus. The template parameters of the plus class are the characteristic bit sequences of the operands of the addition. The code for the plus class is:

```
template<unsigned int N, unsigned int M> class plus {
public:
    static inline void
    add(const CTSV<N>& a, const CTSV<M>& b, CTSV<N|M>& c) {
        add(a.head, b.head, c.head);
        plus< N>>1, M>>1 >::add(a.GetTail(), b.GetTail(), c.GetTail() );
    }
};
template<> class plus<0,0> {
public:
    static inline void add(const CTSV<0>& a, const CTSV<0>& b,
    CTSV<0>& c){}
};
```

In the body of the operator+, an object of the resulting CTSV type is created and passed to the function plus<N,M>::add along with the vectors to be added. This function adds the heads of the vectors with the add function of the Elem<N> classes (defined below) and calls the plus<N>>1,M>>1 >::add function recursively with the tails of the operands. The template parameters are shifted right during the recursion, leading eventually to a call to plus<0,0>::add, which ends the recursion. The add functions for the Elem classes are:

inline void add(const Elem<0>& a, const Elem<0>& b, Elem<0>& c){}
inline void add(const Elem<0>& a, const Elem<1>& b, Elem<1>& c){c.v=b.v;}
inline void add(const Elem<1>& a, const Elem<0>& b, Elem<1>& c){c.v=a.v;}
inline void add(const Elem<1>& a, const Elem<1>& b, Elem<1>& c)
            {c.v=a.v+b.v;}

The resulting type is "promoted" from Elem<0> to Elem<1> if an Elem<1> type object is involved in the operation. In this way the types are handled correctly. It is easy to see that the compilation of these definitions yields optimal code: for addition of two Elem<0> objects, no code is produced, the addition of Elem<0> and Elem<1> results in a single move, and the addition of two Elem<1> objects generates a single addition.

2.2 Generated Code

To clarify how the above works, we examine the previous example more closely. Vectors $x = (< 1, 1 >, < 1, 3 >)$ and $y = (< 2, 3 >, < 2, 5 >)$ can be constructed and added with the code: CTSV<5> x(1); CTSV<20> y(2); x+y; Note that $5 = 00101_2$ and $20 = 10100_2$. Compilation[1] with Borland C++ 5.01 for Intel 80486 using no optimisation resulted the assembly language code:

```
mov dword ptr [ebp-12],1 // create x       mov ecx,dword ptr [ebp-169] // 5th
mov dword ptr [ebp-7],1                     mov dword ptr [ebp-182],ecx
mov dword ptr [ebp-174],2 // create y       mov eax,dword ptr [ebp-192] // ret
mov dword ptr [ebp-169],2                   mov dword ptr [ebp-208],eax
mov eax,dword ptr [ebp-12] // x+y: 1st      mov edx,dword ptr [ebp-187]
mov dword ptr [ebp-192],eax                 mov dword ptr [ebp-203],edx
mov edx,dword ptr [ebp-7] // 3rd            mov ecx,dword ptr [ebp-182]
add edx,dword ptr [ebp-174]                 mov dword ptr [ebp-198],ecx
mov dword ptr [ebp-187],edx
```

The initialisations are the two inevitable move commands for x an y, both having two nonzero elements. Two moves (1st and 5th position) and one addition (3rd) are needed to carry out x + y, which can be seen from the resulting code. The last six lines of the assembly language code originate from the return statement and the implicit invocation of the copy constructor. A clever compiler may avoid this by creating the object directly on the caller's stack. Even without a compiler supporting this *named return-value optimisation* technique, the extra copy can be avoided if we content ourselves with a bit more awkward syntax and use the add function of the plus class directly.

---

[1] Template operator+ was instantiated manually due to limited template support.

3 An Application: Automatic Differentiation

As an alternative to symbolic calculation of derivatives or using approximate difference values, *automatic differentiation* can be used to obtain derivatives. The derivatives are computed using the well-known chain rule. The function value and the derivatives are evaluated simultaneously with the same expression, but instead of scalars we compute using *automatically differentiable numbers* (ADN). These are objects consisting of a function value and a vector of partial derivatives at the same point. To implement the method, we code the differentiation rules for elementary mathematical operations. In C++ this is done by overloading these operations for ADN objects. There are several texts describing automatic differentiation [2, 5] and also software packages available [6, 7].

The derivative vectors of ADNs are usually sparse. Typically there is only one nonzero derivative at the leaves of an expression tree. When approaching the root, the derivative vectors become more dense. CTSVs are ideally suited for ADN derivative vectors.

A class definition for ADNs using CTSVs as the derivative vectors can be easily constructed. Two data members are needed: a floating-point number for the value of the ADN, and a CTSV for the derivatives. So an ADN class is also generic and shares the template parameter of the derivative CTSV. The overloading of elementary mathematical functions and operators for ADN objects is also simple, with vector addition and scalar multiplication of CTSVs defined. To use ADNs in expressions, a certain element position i is chosen for each variable. Then the characteristic bit sequences having only the ith bit set corresponds to the ith variable. The expression is written using ADN objects with these bit sequences: ADN<1>, ADN<2>, ADN<4> and so forth.

Since CTSVs have no run-time penalty for indexing, ADN expression calculations can be further improved. At the leaf level of the expression tree, the derivatives are either 0 or 1, leading to many multiplications by 1. These can be avoided by keeping track of the positions of 1's. Instead of having zero and nonzero elements, we then have zero, unity and arbitrary elements. There is of course no point in tracking the unity elements at run time just to replace a few multiplications with value moves. But since with CTSVs the tracking is done at compile time, it is perfectly feasible and will in some cases produce significant savings in computation time. Since we now distinguish between three types of elements, there must be three specialisations of the Elem class, so two bits of the characteristic bit sequence of the CTSV are needed for each element. See [4] for C++ code and a more detailed discussion about CTSVs in automatic differentiation.

4 Test Runs

The speed of CTSV operations was compared with ordinary dense vectors (regular C arrays) and sparse vectors with indexing performed at run time (vectors of value/index pairs). The speeds of the addition operations were measured with

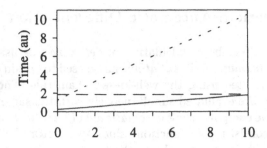

Fig. 1. CPU time of vector addition. *Solid line:* CTSVs; *Dashed line:* dense vectors; *Dotted line:* run-time sparse vectors. Number of nonzero elements is on the x-axis.

10-element vectors, and the number of nonzero elements ranged from 0 to 10. The code was compiled with Borland C++ 5.01 for Intel Pentium processor and optimised for speed. Since the optimiser could not perform return-value optimisation, we used the more awkward syntax to avoid the extra copy constructor invocations. The sparse vectors were allocated statically. Consequently the run-time penalty originates only from indexing, not from memory allocation. The addition functions were called in a loop, and the operand parameters were passed by reference. The CTSV addition for 0 nonzero elements yielded no code. Hence the cost of the function call and looping should approximately equal the cost of the CTSV<0> case. The results are shown in Fig. 1.

The speed of ADN expressions was compared to manually optimised symbolically differentiated code. The ADN derivative vectors were implemented as CTSVs (using the refinement described above), as ordinary dense vectors and as ordinary sparse vectors. Results of the test runs are shown in Fig. 2.

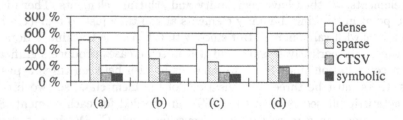

Fig. 2. Relative CPU time (symbolic = 100%) for computing expressions and their derivatives. (a,b): product of five variables differentiated with respect to all variables. (c,d): $\sum_{i=1}^{3} a_i e^{b_i t}$ differentiated with respect to a_i and b_i. (a,c): double variables. (b,d): complex variables.

5 Discussion

A collection of classes for representing sparse vectors in C++ was described. The sparseness pattern of the vector must be known at compile time. For special applications where this restrictive precondition is met, very efficient code can be generated. It was shown how common vector operators can be overloaded for the presented CTSV classes to generate this minimal code from abstract vector expressions.

The classes rely heavily on C++ templates, especially on recursive definitions with non-type template parameters. The sparseness pattern of each vector is represented as a template parameter, i.e. as part of the type information. The sparseness pattern changes in vector operations. Each operation may potentially produce a new vector type. These new template instances are automatically generated from the vector templates by the compiler when encountered. Some of the template features used are quite new and not yet available at all compilers. The features are however part of the standard proposal for C++ [3].

Execution of selected vector operations was compared with dense vectors and ordinary sparse vectors. The speed of CTSVs outperforms both alternatives. The execution speed depends to some extent on the compilers optimisation capabilities; for best results the compiler should be capable of return-value optimisation.

Automatic differentiation was presented as an application of CTSVs. It was shown how to define template classes for automatically differentiable numbers using CTSVs as the derivative vectors. Speed of evaluation for some common expressions was compared with ADNs having alternative vector representations. With our examples, ADNs with CTSV derivatives required CPU time ranging from 20% to 50% of the time used by ADNs with alternative vector representations. The execution speed was only slightly slower than the speed of the symbolically differentiated code.

References

1. Veldhuizen, T.: Using C++ template metaprograms. C++ Report **7** (1995) 36-43.
2. Rall, L. B.: Automatic Differentiation: Techniques and Applications. Lecture Notes in Computer Science **120** (1981) Springer-Verlag, Berlin.
3. 1997 C++ Public Review Document: Working Paper for Draft Proposed International Standard for Information Systems – Programming Language C++. ANSI X3J16/96-0225 ISO WG21/N1043.
4. Järvi J.: Compile Time Sparse Vectors in C++. Turku Centre for Computer Science Technical Report No. 107 (1997) (http://www.tucs.abo.fi/publications).
5. Barton, J. J., Nackman L.R.: Scientific and Engineering C++. Ch. 19, Addison-Wesley, Reading Massaschusetts, 1994.
6. Griewank, A., Juedes, D., Utke, J.: ADOL-C: A Package for the Automatic Differentiation of Algorithms Written in C/C++. ACM Transactions on Mathematical Software **22** (1996) 131–167.
7. Bischof, C. H., Carle, A., Corliss, G. F., Griewank, A., Hovland, P.: ADIFOR: Generating derivative codes from Fortran programs. Sci. Prog. **1** (1992) 1-29.

Will C++ Be Faster Than Fortran?

Todd L. Veldhuizen and M. Ed Jernigan

University of Waterloo, Canada

1 Introduction

As of 1994, the performance of C++ for scientific computing was disappointing. Typical benchmarks showed C++ lagging behind Fortran's performance by 20% to a factor of ten. Performance was so poor that mixed-language programming was necessary: a program's framework could be written in C++, but any speed-critical routines had to be coded in Fortran.

In the past five years there have been significant developments. There are now C++ compilers which perform C++ specific optimizations, such as KAI C++ (from Kuck and Associates Inc.) and Intel C++. Two new programming techniques (expression templates and template metaprograms) perform optimizations such as loop fusion and algorithm specialization. Recent benchmarks show C++ encroaching steadily on Fortran's high-performance monopoly. The latest results are so encouraging that we can pose a heretical question: Will C++ be *faster* than Fortran?

We'll start by looking at reasons why C++ used to be slower, and then describe techniques which can make it faster than Fortran. Ideas will be illustrated with examples and benchmarks from the Blitz++ class library.

2 Why was C++ slower?

2.1 Pairwise expression evaluation

Operator overloading in C++ permits a natural notation for array operations. For example, one can write z = w + x + y; where w, x, y, and z are vectors. Unfortunately, overloaded operators in C++ are always evaluated in a pairwise manner. In a straightforward implementation, evaluation of z = w + x + y; will result in two temporary vectors and three loops being generated. For large vectors, speeds of about 40% Fortran can be achieved (depending on the expression). For tiny vectors ($N < 10$), the cost of allocating temporaries dominates, and the speed is 5-10% Fortran speed.

This problem has been solved by the expression templates technique [6]. Expression templates are used to build parse trees of expressions at compile time; the parse trees are encoded as type names. Once the parse tree has been encoded, it can be manipulated in interesting ways. One application of expression templates solves the pairwise expression evaluation problem: code such as z = w + x + y; is transformed into

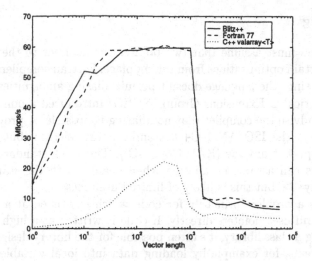

Fig. 1. Results for the DAXPY benchmark

```
for (int i=0; i < N; ++i)
    z[i] = w[i] + x[i] + y[i];
```

Expression templates have some overhead associated with constructing the parse tree. However, KAI C++ optimizes away this overhead completely, resulting in code which is just as fast as Fortran 77.

Figure 1 shows benchmark results for a DAXPY operation implemented using the Blitz++ library, Fortran 77 and the `valarray<T>` class from the ISO/ANSI C++ library.[1] The `valarray<T>` performance is typical of a pre-expression templates implementation. The sharp performance drop around N=2000 is due to cache effects.

Although Fortran 90 supports operator overloading, there's no comparable mechanism to expression templates. This limits the generation of efficient kernels by F90 compilers to expressions involving arrays which are dense, asymmetric, box-shaped, and linearly-addressable. The expression templates mechanism in C++ has no such limitations: it can generate efficient kernels for sparse, symmetric, jagged-edged, and/or nonlinearly-addressed arrays.

---

[1] All benchmark results were measured on a 100 MHz RS/6000 Model 43P using KAI C++ 3.2c at +K3 -O3, XL Fortran 77 3.2.5 at -O3, and XL Fortran 90 3.2.5 at -O3. The XL Fortran 90 compiler was omitted from the DAXPY benchmark because it performed poorly. DAXPY is the Level 1 BLAS which implements the vector operation `y += a * x`.

2.2 Aliasing

C++ compilers must assume that two pointers might refer to the same data, preventing certain optimizations from taking place. Fortran compilers don't have this problem, since the language doesn't permit aliasing ambiguities.

The Numerical C Extensions Group (NCEG) introduced a pointer qualifier restrict to advise the compiler that no aliasing is possible. A proposal to add this keyword to the ISO/ANSI C++ standard was rejected, but some C++ compilers support it anyway (KAI, Cray, SGi). The C++ standard does state that compilers can assume no aliasing when dealing with the standard array class valarray<T>, but this is likely of limited usefulness.

Aliasing is a problem primarily for code which is written at a low level of abstraction, and uses pointers directly. If code is written at a high level of abstraction using a class library, it's often possible for the library designer to avoid aliasing problems, for example by loading data into local variables which are guaranteed alias-free.

3 Why might C++ be faster?

3.1 Abstraction can make optimization easier

Optimization of low-level languages such as Fortran 77 requires a transformational approach: a compiler must first understand the intent of the program, and then transform it into a faster (but equivalent) program. Optimizing code written at a high level of abstraction is often easier, since a generative (rather than transformational) approach is possible. Abstract programs resemble specifications more than implementations. Complex, machine-specific code can be generated to carry out the operation.

In Fortran 90/95, the array abstraction is built into the language; the compiler generates optimized low-level code to implement the high-level program. C++ has an advantage because it provides library designers with the tools they need to build their own abstractions; optimized code can be generated by the library itself using template techniques.

Consider this code excerpt which implements the 2-D acoustic wave equation using Blitz++ arrays:

```
Array<float,2> P1(N,N), P2(N,N), P3(N,N), c(N,N);   // ...
Range I(1,N-2), J(1,N-2);

for (int iter=0; iter < niters; ++iter) {
    P3(I,J) = (2-4*c(I,J)) * P2(I,J)
        + c(I,J)*(P2(I-1,J) + P2(I+1,J) + P2(I,J-1) + P2(I,J+1))
        - P1(I,J);
    P1 = P2;
    P2 = P3;
}
```

The performance of this stencil operation is constrained by memory bandwidth. Three optimizations are required for good performance: tiling, partial unrolling, and avoiding the array copy operations. Blitz++ performs the first two optimizations automatically, and the third is achieved by a single line of code. It's possible to do the same optimizations in Fortran 77 and 90, but the clarity of the code suffers.

	Time	Mflops/s	Code
Untuned versions			statements
Fortran 90	93.3 s	9.7	12
Fortran 77	53.2 s	17.0	21
Blitz++	57.8 s	15.7	9
Older C++ math library	418.2 s	2.1	9
Tuned versions			
Fortran 90	41.1 s	22.1	19
Fortran 77	27.3 s	33.2	39
Blitz++	31.6 s	28.7	8
Older C++ math library	361.9 s	2.4	11

Table 1. 2D Acoustic benchmark results for 240 iterations on a 650^2 grid.

Table 1 shows benchmark results for this problem[2]. The Blitz++ versions are highly expressive (very few lines of code are required to implement the benchmark) and compete with the performance of the Fortran versions. The tuned Blitz++ and Fortran 77 implementations do exactly the same optimizations. The tuned Fortran 90 version avoids array copying, but leaves the traversal order up to the compiler. Although the Fortran 77 version is slightly faster than Blitz++, it took 30 lines of source code and several days of painstaking tuning to achieve this result, and the code is not reusable for other stencils. The Blitz++ library generates an equivalent implementation for any array stencil, triggered by a single line of source code. Moreover, the code which it generates can be platform specific.

It is possible to do better than the tiled traversal, by using an iteration space tiling (IST) (see e.g. [9]). Such tilings are fiendishly complicated to implement, and are currently far beyond the ability of optimizing compilers. In C++, it's possible to create template-based frameworks for ISTs. Such a framework for finite-differenced PDEs would require users to provide their stencil operation and boundary conditions. These would be combined with the generic IST code to produce a near-optimal implementation. It's not possible to implement such frameworks in Fortran 77/90/95, since these languages lack generic programming features.

In addition to generating special cache-optimized traversals for 2D and 3D stencil operations, Blitz++ does other loop transformations which previously had been the sole domain of optimizing compilers, such as loop interchange, hoisting stride calculations out of inner loops, partial unrolling, collapsing inner

---

[2] Source code available: see http://monet.uwaterloo.ca/blitz/.

loops, and optimizations for unit strides. Features "on the drawing board" include automatic padding of arrays to avoid cache interference, and loop fusion for multiple array expressions.

3.2 Directed algorithm specialization

It's often possible to speed up an algorithm by specializing it for a particular situation. Scientific computing algorithms are known to benefit greatly from specialization [1]. Consider this dot product algorithm:

```
double dot(double* a, double* b, int n) {
    double result = 0.0;
    for (int i=0; i < n; ++i)
        result += a[i] * b[i];
    return result;
}
```

If we wanted to use dot() for vectors of length 3, the performance would be a fraction of peak performance due to function call and looping overhead. A specialized version eliminates the performance loss:

```
inline double dot3(double* a, double* b) {
    return a[0] * b[0] + a[1] * b[1] + a[2] * b[2];
}
```

There are two competing approaches to generating specialized algorithms: partial evaluation and generative metaprogramming. Partial evaluators use a transformational approach to optimization: they take a program, analyze it, and transform selected algorithms into specialized versions. Many compilers perform very limited forms of partial evaluation.

Generative metaprogramming provides an alternative to partial evaluation. The term implies programs which generate source code for specialized algorithms. The technique is well-known among computational scientists: examples include producing source code for fixed-size, unrolled Fast Fourier Transforms and matrix multiplication code for matrices with known sparse structure. It turns out that C++ provides crude generative metaprogramming facilities quite by accident, as a byproduct of its template instantiation mechanism [7]. These *template metaprograms* are capable of doing arbitrary computations at compile time, and produce compilable C++ code as their output.

Template metaprograms can be used to specialize algorithms. For example, an FFT implemented using template metaprograms can calculate the appropriate complex roots of 1 and array shuffle at compile time, unroll all the loops, and output a solid block of floating-point math code. An important distinction between template metaprograms and partial evaluation is that the algorithm specializations produced by template metaprograms are unlimited in complexity and *directed* – one has complete control over exactly which optimizations are performed.

A prime candidate area for algorithm specialization in scientific computing is manipulation of small, dense vectors and matrices [8]. Consider the equation $y = x - 2(x \cdot n)n$ which reflects an incident ray x off a surface with normal n. A Blitz++ implementation of this equation would use the `TinyVector<T,N>` class:

```
typedef TinyVector<double,3> ray;

void reflect(ray& y, const ray& x, const ray& n) {
    y = x - 2 * dot(x,n) * n;
}
```

The `TinyVector<T,N>` class places the elements of the vector directly on the stack, so there is no memory allocation overhead. Using a combination of template metaprograms and expression templates, the above code is transformed into

```
double _t1 = x[0] * n[0] + x[1] * n[1] + x[2] * n[2];
double _t2 = _t1 + _t1;
y[0] = x[0] - _t2 * n[0];
y[1] = x[1] - _t2 * n[1];
y[2] = x[2] - _t2 * n[2];
```

The back-end compiler then reorders this code for superscalar execution, and produces a mere 20 instructions. Specializing algorithms for small vectors and matrices exposes low-level parallelism, eliminates function call overhead, permits data to be registerized, and allows floating point operations to be scheduled around adjacent computations. The resulting algorithms are up to 10 times faster than calls to library subroutines.

3.3 Encapsulation can improve locality of reference

Encapsulation is a basic idea of object-oriented design: you take data which is closely related, and package it as an object. A side benefit of encapsulation is that it encourages locality of reference, resulting in favorable memory access patterns [5]. This point is nicely illustrated by a lattice quantum chromodynamics benchmark based on an implementation from the Edinburgh Parallel Computing Centre (EPCC) [2]. In the EPCC implementation, much CPU time was consumed by multiplication of 2-spinors by $SU(3)$ gauge elements; this is equivalent to multiplying 3x2 and 3x3 complex matrices. Using the Blitz++ library, one can use the class `TinyMatrix<T,N,M>` for the spinors and gauge elements. This class uses template metaprograms to create specialized algorithms for matrix multiplication. The initial version of the benchmark was this:

```
typedef TinyMatrix<complex<double>,3,3> gaugeElement;
typedef TinyMatrix<complex<double>,3,2> spinor;

void qcdCalc(Vector<spinor>& res, Vector<gaugeElement>& M,
    Vector<spinor>& src)
```

```
{
    for (int i=0; i < res.length(); ++i)
        res[i] = product(M[i], src[i]);
}
```

To tune this version, the gauge element and related spinors were encapsulated as a single object:

```
class latticeUnit {
...
protected:
    spinor one;
    gaugeElement ge;
    spinor two;
};
```

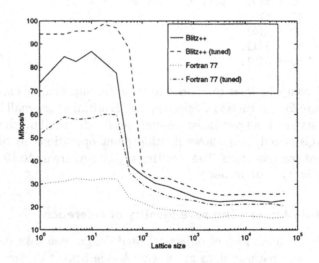

Fig. 2. Results for the lattice QCD benchmark

Figure 2 shows the resulting performance for the initial and tuned Blitz++ versions, as well as a Fortran 77 version. Encapsulation improved the performance by about 12% in the out-of-cache region. Despite extensive hand-tuning of the Fortran version (experimenting with loop-unrolling combinations), it was unable to achieve the performance of either C++ versions due to register spillage.

4 Concluding remarks

C++ is now ready to compete with the performance of Fortran. Its performance problems have been solved by a combination of better optimizing compilers (such as KAI C++) and template techniques. It's possible that C++ will be faster than Fortran for some applications, since:

- Fortran 90 has the array abstraction built into the language. This limits the ability of the compiler to generate efficient code to expressions involving simple (e.g. dense, asymmetric) arrays. The expression templates technique in C++ has no such limitations; it can be applied to more complicated (e.g. sparse, jagged-edged) arrays.
- Template metaprograms can generate specialized algorithms well beyond the ability of current Fortran compilers.
- C++ templates allow the creation of generic frameworks for common scientific computing tasks (such as solving finite-differenced PDEs). Such frameworks can implement complex optimizations (such as iteration-space tilings).
- The object-oriented nature of C++ encourages favourable memory access patterns.

5 Acknowledgments

This work was supported in part by the National Science and Engineering Research Council of Canada, and by Rogue Wave Software of Corvallis, Oregon. Axel Thimm suggested useful improvements to the QCD example. Arch Robison and colleagues at KAI provided helpful advice on the use of their compiler.

References

1. Andrew Berlin and Daniel Weise. Compiling scientific code using partial evaluation. *Computer*, 23(12):25–37, Dec 1990.
2. Stephen Booth. Lattice QCD simulation programs on the Cray T3D. Technical Report EPCC-TR96-03, Edinburgh Parallel Computing Centre, 1996.
3. Scott Haney. Is C++ fast enough for scientific computing? *Computers In Physics*, 8(6):690–694, Nov/Dec 1994.
4. Rebecca Parsons and Daniel Quinlan. A++/P++ array classes for architecture independent finite difference computations. In *Proceedings of the Second Annual Object-Oriented Numerics Conference (OON-SKI'94)*, pages 408–418, April 24–27, 1994.
5. Arch D. Robison. C++ gets faster for scientific computing. *Computers in Physics*, 10(5):458–462, Sep/Oct 1996.
6. Todd Veldhuizen. Expression templates. *C++ Report*, 7(5):26–31, June 1995. Reprinted in C++ Gems, ed. Stanley Lippman.
7. Todd Veldhuizen. Using C++ template metaprograms. *C++ Report*, 7(4):36–43, May 1995. Reprinted in C++ Gems, ed. Stanley Lippman.
8. Todd Veldhuizen and Kumaraswamy Ponnambalam. Linear algebra with C++ template metaprograms. *Dr. Dobb's Journal of Software Tools*, 21(8):38–44, August 1996.
9. Michael Wolfe. Iteration space tiling for memory hierarchies. In Gary Rodrigue, editor, *Proceedings of the 3rd Conference on Parallel Processing for Scientific Computing*, pages 357–361, Philadelphia, PA, USA, December 1989. SIAM Publishers.

The Design and Evolution of the MPI-2 C++ Interface

Jeffrey M. Squyres[1], Bill Saphir[2], and Andrew Lumsdaine[1]

[1] Department of Computer Science and Engineering, Notre Dame, IN 46656
[2] NERSC, Lawrence Berkeley National Laboratory, Berkeley, CA 94720

Abstract. The original specification for the Message Passing Interface (MPI) included language bindings for C and Fortran 77. C++ programs that used MPI were thus required to use the C bindings. With MPI-2, a C++ interface for all of MPI is specified. In this paper, we describe the design of the C++ interface for MPI and provide some of the history and motivations behind the design decisions.

1 Introduction

The Message Passing Interface (MPI) [2, 4, 6] is a specification for a library of routines that provide an infrastructure for parallel message passing applications. MPI provides routines for point-to-point communication, collective operations, as well as support for the development of safe libraries and other related functionality. The MPI standard defines C and Fortran (77) bindings for all MPI functions.

The MPI Forum reconvened in 1995 to consider additions to the MPI standard, known as MPI-2. MPI-2 has now been finalized [3]. MPI-2 defines a C++ interface for all MPI-1 and MPI-2 functions.

The development of a C++ interface followed a winding path as the Forum considered many different styles of interfaces. What eventually emerged is closely related to the C interface, but has a number of important features that make it more appealing to C++ programmers and exploit MPI features in new ways.

In this paper we describe the design of the MPI-2 C++ interface, explaining what choices were made and why. We assume that the reader is familiar with MPI in general and the MPI C bindings in particular.

2 The Major Issues

2.1 Big or Small?

A number of proposals for the MPI C++ bindings were introduced during the course of the MPI-2 Forum. The original (preliminary) proposal was modeled closely after the MPI++ [5] class library. The initial proposal introduced a major question to the Forum: Should the bindings be a full-blown class library or should they be something closer to the C interface? Both options were explored, with proposals for each being made over a period of time. After the Forum had a chance to study and evaluate the class library proposal, it was felt that the role of the C++ bindings was to facilitate the development of class libraries, not to actually be a class library. The proposed class library later became Object-Oriented MPI [7].

After the class library approach was discarded, the pendulum swung the other way and a proposal for very low-level bindings was made. These proposed bindings were very close to the C bindings, but provided a few C++ features such as const and reference semantics. However, the Forum felt that these bindings were too low-level and did not do enough to enable class library design.

Thus, the final, and accepted, proposal for MPI C++ bindings found the middle ground between big and small (or, like MPI itself, it was *both* big and small). The bindings contain a number of class library-like features, but still remain limited enough not to constrain class libraries built using them.

2.2 Object-Based Design

The design of MPI itself is object-based. MPI defines a number of objects — Communicators, Groups, Requests, etc. These objects are referred to by handles in C and by integers in Fortran. It was an obvious choice for the C++ bindings (once there was a decision to go with the "small" interface) to turn the handles into regular C++ objects. These objects, however, retained the same handle-based semantics as their C counterparts. Namely, the C++ objects are user-level handles to underlying implementation-dependent objects.

Most MPI functions became methods associated with these objects. In most cases, which object to associate with a given function was "obvious" to the Forum, though the rationale is ultimately more intuitive than rigorous. Examples of these obvious choices were Comm.Send() and Request.Wait(). There are good arguments for preferring Comm.Send() to Datatype.Send(), for instance, but this was not an issue for the Forum, and we do not discuss them here.

Some functions were "obvious" candidates for a specific class even though they did not include a single *IN* or *OUT* argument of that type. For example, MPI::Datatype::Create_struct takes an *IN* array of MPI::Datatype. Such functions were still defined on that class, but were declared static since they have no corresponding this pointer.

2.3 Naming

MPI-1 did not use consistent naming rules. Often, names are of the form MPI-_Object_action as in MPI_COMM_SPLIT and MPI_INTERCOMM_MERGE, but sometimes they are not, e.g., MPI_TYPE_CONTIGUOUS. Sometimes the verbs are consistent, e.g., MPI_COMM_FREE and MPI_TYPE_FREE, but sometimes they are not, as in MPI_ERRHANDLER_SET and MPI_ATTR_PUT.

Unlike MPI-1, MPI-2 uses a consistent naming scheme of the form MPI-_Object_action_subset. For the C++ bindings, the Forum decided to use the consistent names. Although the MPI-2 C++ scheme for both MPI-1 and MPI-2 functions are slightly different from the MPI-1 C names in several cases, the consistent C++ naming scheme was felt to be advantageous for the following reasons:

1. The C++ bindings are new. There is no existing code that needs to be changed or book that must be rewritten.
2. It was felt that using the inconsistent names was more of a problem in a C++ context where the structure highlights discordant design. The inconsistent names would result in C++ names such as Status.Get_count() and Status.Get_elements(), which both use the verb "get," while Comm.Size() and Comm.Rank() do not.[1]
3. The C++ names are necessarily different from the C names already, e.g., Comm.Send() instead of Comm.MPI_Send().

One relatively new feature of ANSI C++ is the namespace construct which allows programs to provide explicit scoping of MPI names. The MPI C++ bindings make use of this feature by including all names within the scope of a namespace MPI.[2] As such, all C++ MPI names are prefixed with "MPI::".

2.4 Object Semantics

Constructors and Destructors. MPI-1 has routines that clearly create objects (e.g., MPI_COMM_DUP) and routines that free them (e.g., MPI_COMM_FREE). It seems at first natural in C++ to turn these and related functions into constructors and destructors. The main problem with such an approach is that Create() and Free() are collective operations. Thus a declaration

    MPI::Comm a(MPI::COMM_WORLD)

intended to implicitly MPI_COMM_DUP the predefined MPI_COMM_WORLD communicator would be a "collective declaration." Worse, the return from the routine where this variable was declared would be a collective operation, when the object was implicitly freed with MPI_COMM_FREE in the destructor.

MPI-2 therefore chooses a path that may be unfamiliar to C++ programmers: the application is responsible for explicitly creating and freeing objects using the appropriate explicit MPI function calls. Consitent with the MPI memory management model, memory management is not automatically handled by constructors and destructors.

As a consolation prize, MPI-2 specifies default constructors that initialize objects to be equivalent to their corresponding MPI::*_NULL handles, and destructors that do free the "top-level" C++ object, but not the underlying object to which it refers.

Copy and Assignment. Since the C++ objects are still handles to underlying objects, the copy and assignment operations are shallow. The assignment

---

[1] In comparison, the actual names of the last two functions use the standardized verb "get". They are MPI::Comm::Get_size() and MPI::Comm::Get_rank(), respectively.
[2] Some C++ compilers do not implement the namespace construct yet. An *Advice to implementors* in the MPI-2 standard allows implementors to use a non-instantiable MPI class if namespace is not available.

```
MPI::Comm comm = MPI::COMM_WORLD;
```

does not create a new communicator, comm is now an alias for MPI::COMM_WORLD. That is, comm and MPI::COMM_WORLD now reference the same underlying object. The MPI::Status object does not follow this rule. Since MPI::Status has public data members and does not necessarily point to internal implementation-dependant data, copies and assignments are deep.

Comparison. Similarly, comparisons of MPI handles return true only if they point to the same internal object. MPI::Status is again an exception to this rule; the comparison operators are not defined on the MPI::Status class because it is not a handle to an underlying object.

Constants. The predefined MPI constants are singleton objects, meaning that they can only be instantiated once. In contrast to the C and Fortran constants, their types are explicitly specified with the exception of MPI::COMM_NULL (discussed below). All constants must also be const to allow for compiler optimizations, particularly when passed as function parameters.

2.5 The Comm Class Hierarchy

There are four types of communicators in MPI: inter-, intra-, Cartesian, and graph. These distinct types are represented by a single type in C and Fortran, i.e., MPI_COMM. However, many functions that take an MPI_COMM argument only take a specific *kind* of MPI_COMM argument, e.g., MPI_CART_RANK and MPI_GRAPH_NEIGHBORS take a Cartesian and graph communicator, respectively. Passing in the wrong kind to these functions results in a run-time error.

C++ provides a natural way to provide better type checking of communicators while still keeping the communicator types under the MPI::Comm umbrella, i.e., deriving new types from the abstract base class MPI::Comm. Thus, there are four communicator classes: Intercomm, Intracomm, Cartcomm, and Graphcomm. Intercomm and Intracomm are derived from the Comm base class; Cartcomm and Graphcomm are derived from Intracomm. Functions that require specific types of communicators (e.g., MPI_CART_RANK) are defined on their respective classes, while functions that apply to all types of communicators (e.g., MPI_SEND) are defined on the base class, MPI::Comm.

The function MPI_COMM_DUP presents a unique problem in that MPI_COMM_DUP returns a new communicator of the same type as the original. In C, this is not a problem because all four communicators are of the same type. In C++, however, each communicator is a different type, and one function cannot return four different types. Also, since it is not desirable to have copy constructors that perform collective actions, DUP must be realized as a regular member function. Several alternatives were proposed:

1. *Return by Reference.* virtual MPI::Comm& MPI::Comm::Dup() This binding does not fit the MPI memory management model that requires the user to manage memory. Thus, this C++ version of MPI_COMM_DUP would be significantly different than its C and Fortran counterparts if it were to allocate memory itself.

2. *Return by* INOUT. `virtual void MPI::Comm::Dup(MPI::Comm& newcomm)`
Although this binding would not break the MPI memory management model,
the syntax is non-intuitive, and does not conform to ideas discussed in Section 2.7

3. *Return by value.* `COMMTYPE MPI::COMMTYPE::Dup()` Substituting any of the
four communicator types for `COMMTYPE` gives four `non-virtual` bindings; this
function is not implemented on the `MPI::Comm` base class. However, typical
parallel library functions take any type of communicator as an argument,
duplicate it (regardless of its type), and use it to perform simple sends and
receives. That is, a typical function in a C++ parallel library may be pro-
totyped as:

    void startFoo(Comm& usercomm)

But since `Dup()` is not defined on the `Comm` base class, `startFoo` cannot
duplicate `usercomm` without first casting it to another type, which defeats
the point of prototyping the argument as a `Comm&`. Simply put, this binding
restricts the use of ploymorphism.

The Forum decided that option 3 was the best solution since it does not
break the MPI memory manangement model, has intuitive syntax, and returns
the correct type. To provide a "virtual `Dup()`", a new function was introduced
that only exists in the C++ bindings, `Clone()`:

    virtual MPI::Comm& MPI::Comm::Clone() = 0;

Although `Clone()` does not conform to the MPI memory management model,
the Forum considered that its lack of symmetry is acceptable because the name
only exists in the C++ bindings. Note: that the prototypes of `Clone()` in the
derived classes are slightly different; they return references to their respective
(derived) types.[3]

The type of `MPI::COMM_NULL` is implementation dependent; it must be able
to be used in comparisons and initializations with all types of communicators.

2.6 Exceptions

The C bindings for almost all MPI functions (except `MPI_WTICK` and `MPI_WTIME`)
return an error code. In principle, an application can check this error code and
take some action if there is an error. In practice, this error code is rarely used.
First, by default, errors cause an MPI program to abort (this is often the desired
behavior). Second, even if MPI is configured to return errors to the application,
the MPI standard states that the state of a program is undefined after an MPI
error. About the only thing an application can (semi)reliably do is print and
error message and abort. Finally, it is tedious to check the return value from
every MPI function call and to appropriately handle the errors.

---

[3] Not all C++ compilers implement `virtual` functions in derived classes that can
overload the return types. An *Advice to Implementors* in MPI-2 allows implementors
to return `MPI::Comm&` if their compiler does not yet support this feature.

C++ exception handling provides an elegant mechanism to handle errors. C++ applications are given the option of setting the default error handler to MPI::ERRORS_THROW_EXCEPTIONS, in which case MPI functions throw a C++ exception when there is an error. Thus, C++ methods do not return error codes as function values.

2.7 Return Values

In C and in Fortran, values are returned through the argument list. For instance,

```
MPI_Comm newcomm;
MPI_Comm_dup(MPI_COMM_WORLD, &newcomm);
```

returns a new communicator in newcomm. Part of the reason for this is that the return value of the function is reserved for the error code. Since C++ MPI methods do not return error codes, the function return value is freed up to hold more than an error code, thus allowing more natural notation such as

```
MPI::Comm newcomm = MPI::COMM_WORLD.Dup();
```

In many MPI functions there is a single "*OUT*" quantity that makes sense as a return value in the C++ case. In other functions, the *OUT* quantity may not be readily returned (e.g., when it is an array – because of MPI's memory management model – or when there are multiple *OUT* arguments and it is not obvious which argument should be returned), or there may be no *OUT* quantity at all. In these cases, the corresponding C++ bindings return void.

2.8 References, Pointers, and MPI_STATUS_IGNORE

The MPI C++ bindings use const and reference semantics when possible. All "*IN*" parameters that are MPI objects are both const and passed by reference to allow for compiler optimization. Additionally, passing by reference does not incur the additional overhead of copy constructors. Thus, the binding for MPI_COMM_SEND is

```
void Comm::Send(..., const Datatype& datatype, ...)
```

The only pointer arguments are char* arguments for strings (because of convention) and void* arguments for choice buffer arguments.

This introduces a problem with the the new MPI-2 option to ignore a returned MPI_Status by specifying the constant MPI_STATUS_IGNORE for the corresponding *OUT* argument. In C, this constant is an argument of type MPI_Status*, and usually has the value NULL. In C++, the corresponding status argument is passed by reference and therefore must be a valid MPI::Status instance; it is not possible to pass a NULL pointer. Therefore, the C++ bindings take a different route, which is to have two bindings for every function with an *OUT* MPI_Status argument in the language-neutral specification. One binding has a reference to a MPI::Status argument and the other has no MPI::Status argument. The C constant MPI_STATUS_IGNORE has no corresponding constant in C++.

2.9 Interfacing with C

To provide a transparent interface between C and C++, three functions are defined on all C++ MPI classes (except `MPI::Status`): a casting operator to cast C++ objects into C handles, a promotion operator to create C++ objects from C handles, and an assignment operator to allow the assignment of C handles to C++ objects.

There is no mechanism to translate directly from C++ to Fortran. A user must convert a C++ object into a C handle and then use the provided MPI-2 functions to convert it to a valid Fortran handle.

3 Design Details

An abbreviated definition of the MPI namespace and its member classes is as follows:

```
namespace MPI {
  class Comm                             {...};
    class Intracomm : public Comm        {...};
      class Graphcomm : public Intracomm {...};
      class Cartcomm  : public Intracomm {...};
    class Intercomm : public Comm        {...};
  class Datatype                         {...};
  class Errhandler                       {...};
  class Exception                        {...};
  class File                             {...};
  class Group                            {...};
  class Info                             {...};
  class Op                               {...};
  class Request                          {...};
    class Prequest : public Request      {...};
    class Grequest : public Request      {...};
  class Status                           {...};
  class Win                              {...};
};
```

Multiple and virtual inheritance are *not* used in the design. All member functions are `virtual` except those which are `static` (which cannot be `virtual`) and the `MPI_COMM_DUP` variants (which are implemented separately on each class).

4 Example

Figure 1 shows an example program using the C++ bindings. The program sends a simple integer message around a ring of processors five times. Although the code is not significantly different from a corresponding program written in C, it does show the expressive power and more natural notation of the C++ bindings.

```
#include <mpi.h>

int main(int argc, char *argv[])
{
  MPI::Init(argc, argv);
  int msg = 123;

  int rank = MPI::COMM_WORLD.Get_rank();
  int size = MPI::COMM_WORLD.Get_size();
  int to   = (rank + 1) % size;
  int from = (size + rank - 1) % size;

  if (rank == size - 1)
    MPI::COMM_WORLD.Send(&msg, 1, MPI::INT, to, 4);
  for (int i = 0; i < 5; i++) {
    MPI::COMM_WORLD.Recv(&msg, 1, MPI::INT, from, MPI::ANY_TAG);
    MPI::COMM_WORLD.Send(&msg, 1, MPI::INT, to, 4);
  }
  if (rank == 0)
    MPI::COMM_WORLD.Recv(&msg, 1, MPI::INT, from, MPI::ANY_TAG);

  MPI::Finalize();
  return 0;
}
```

Fig. 1. Canonical "ring" program using the C++ bindings.

References

1. Satish Balay, William D. Gropp, Lois Curfman McInnes, and Barry F. Smith. Efficient management of parallelism in object-oriented numerical software libraries. In E. Arge, A. M. Bruaset, and H. P. Langtangen, editors, *Modern Software Tools in Scientific Computing*. Birkhauser, 1997.
2. Nathan E. Doss, William Gropp, Ewing Lusk, and Anthony Skjellum. An initial implementation of MPI. Technical Report MCS-P393-1193, Mathematics and Computer Science Division, Argonne National Laboratory, Argonne, IL 60439, 1993.
3. MPI Forum. MPI-2: Extension to the message passing interface. Technical report, University of Tennessee, July 1997.
4. William Gropp, Ewing Lusk, and Anthony Skjellum. *Using MPI: Portable Parallel Programming with the Message Passing Interface*. MIT Press, 1994.
5. Anthony Skjellum, Ziyang Lu, Purushotham V. Bangalore, and Nathan E. Doss. Explicit parallel programming in C++ based on the message-passing interface (MPI). In Gregory V. Wilson, editor, *Parallel Programming Using C++*. MIT Press, 1996. Also available as MSSU-EIRS-ERC-95-7.
6. Marc Snir, Steve W. Otto, Steve Huss-Lederman, David W. Walker, and Jack Dongarra. *MPI The Complete Reference*. MIT Press, Cambridge, MA, 1996.
7. Jeffrey M. Squyres, Brian C. McCandless, and Andrew Lumsdaine. Object oriented MPI reference. Computer Science and Engineering Technical Report 96-10, University of Notre Dame, 1996.

Efficient Extensible Synchronization in Sather

Jürgen W. Quittek, Boris Weissman

International Computer Science Institute, 1947 Center St., Berkeley, CA 94704
quittek@icsi.berkeley.edu, borisv@icsi.berkeley.edu,

Abstract. Sather, a parallel object-oriented programming language developed at ICSI, offers advanced thread synchronization constructs separating locking mechanism and policies. While a lock management system provides a general locking mechanism, synchronization objects define and implement different extensible policies. Commonly used synchronization objects such as mutual exclusion and reader/writer locks are provided by the standard Sather library. Synchronization objects with more complex semantics can be defined by the user. The conjunctive and disjunctive acquisition of collections of locks and the deadlock detection are distinct features of Sather supported by the locking mechanism.

This paper introduces the Sather synchronization constructs and presents the design and implementation of a lock management runtime system. We argue that a clean, object-oriented design allows us to support sophisticated synchronization policies while preserving efficiency on distributed computing platforms. The system is fully implemented and runs on several platforms including a network of symmetric multiprocessors connected by a fast, user-level, low latency communications network.

1 Introduction

Simple synchronization objects such as mutual exclusion locks or semaphores are under explicit user control in many parallel systems. However, in distributed or fault-tolerant environments runtime synchronization systems may be necessary to deal with added complexity. For example, distributed lock managers are used for access control in recoverable distributed file servers such as VAXcluster [1] or the Highly Available Network File Server [2]. Transactions in complex object-oriented DBMSs [3] and in distributed databases such as the Oracle Parallel Server [4] are also controlled by lock managers. Lock managers are also used to ensure memory consistency in distributed shared memory systems [5, 6].

The Sather runtime is more general than the examples listed above because it supports arbitrary locking policies. File systems use a mutual exclusion locking policy. Databases and distributed shared memory systems can add reader/writer locking policies. In Sather, synchronization objects define their own policies. Consequently, the common lock management runtime mechanism must be independent of the policies, but general enough to realize any policy.

The Arjuna programming System, a fault-tolerant distributed extension of C++ is an example of a system that provides separation between the synchronization mechanism and policies [7]. Another example is PRESTO, a parallel object-oriented environment for C++ [8]. However, in both Arjuna and PRESTO, only single locks can be acquired, while Sather supports atomic locking of sets of locks (conjunctive locking) and alternative locking of sets of locks (disjunctive locking). This functionality impacts overall design decisions for distributed environments.

2 Thread Synchronization in Sather

Sather is a modern parallel object-oriented programming language designed to be safe and efficient. Some major Sather features include strong static typing, dynamic dispatch, multiple inheritance, parametrized classes, iteration abstraction, garbage collection, and exception handling. Sather is available on most Unix platforms as well as on Windows NT. A more detailed language description can be found in [9, 10]. The Sather language specification, language manual, and full Sather distribution are available at http://www.icsi.berkeley.edu/~sather.

Sather supports concurrent programming based on threads. It offers a shared memory model for distributed platforms and allows the explicit placement of threads and objects.

Runtime synchronization is handled by a *lock management system (LMS)* and by *locks*. The LMS provides an advanced locking mechanism that avoids typical deadlocks and starvation. Locks define locking policies such as a mutual exclusion or reader/writer policy. While it is impossible to foresee all synchronization policies that may be required by different applications, Sather provides a library of commonly used ones. Applications that use more sophisticated synchronization patterns can define their own synchronization policies and extend the standard library. The functionality of the LMS itself is well-defined and supported by the Sather runtime.

2.1 The Lock Statement

The Sather lock statement provides a programming language level interface to the LMS. Simple applications of the Sather lock statement include critical sections [11] and are similar to the synchronized block construct in Java [12]. The lock statement also extends and generalizes many other common synchronization concepts.

A lock statement consists of one or more lists of Sather statements called branches. Each branch is associated with a set of locks. The following code example shows different usage patterns for the lock statement including conjunctive and disjunctive locking.

```
-- declare lock variables
lock1,lock2,lock3:$LOCK;
...
-- single lock synchronization
lock when lock1 then <branch 1> end;
...
-- acquire all three locks atomically
lock when lock1,lock2,lock3 then <branch 2> end;
...
```

```
-- Take either branch 3 if lock1 is available or branch 4
-- if lock2 and lock3 are available. Otherwise, take branch 5
lock
    when lock1 then <branch 3>
    when lock2,lock3 then <branch 4>
    else <branch 5>
end;
```

Locks are synchronization objects which can be acquired, held and released by threads. A branch of a lock statement can be taken only if all locks associated with this branch have been acquired. Only a single branch is guaranteed to be executed, even if several branches can be taken. The *else branch* is executed if no other branch can be taken. In the absence of the else branch, the executing thread is blocked until at least one branch becomes available. After a branch has been executed, all held locks are released.

Acquiring a set of locks is an atomic operation. This allows to avoid many typical deadlock situations. The lock statement guarantees fairness with respect to threads and branches. A thread that can acquire a branch will eventually do so, even if this requires blocking other threads. Another aspect of fairness deals with selecting a branch when more than one branch are available. In this case, each branch is taken with a non-zero probability. This avoids starvation of individual branches.

2.2 Synchronization Objects: Locks

While the semantics of the lock statement is well-defined, the locks used in the statement may behave differently. The Sather distribution provides a library of locks including mutual exclusion locks, reader/writer locks and other more complex locks. Locks with other semantics can be defined by the user.

Any lock has to be a subtype of the abstract class $LOCK. In Sather, abstract classes define interfaces and provide no implementation. Any concrete lock class that inherits from $LOCK must fully implement its interface:

```
abstract class $LOCK is
    acquirable(tid:THREAD_ID):BOOL;
    acquire(tid:THREAD_ID);
    release(tid:THREAD_ID);
    request(tid:THREAD_ID);
    cancel_request(tid:THREAD_ID);
    wait_for(tid:THREAD_ID):ARRAY{THREAD_ID};
end;
```

The lock interface methods may be called only by the LMS. They cannot block, raise exceptions or recursively execute other lock statements.

acquirable returns true if the thread identified by tid can acquire this lock. Methods acquire and release notify the lock that it is has been acquired or released respectively. request is executed for each thread blocking on the lock. Method cancel_request is called, when a thread unblocks and can continues execution.

`wait_for` is used for deadlock detection. It returns an array of threads that have to release the lock before it can be acquired by the thread identified by `tid`.

The programmer's freedom to define the semantics of synchronization objects enables the customization of synchronization policies. For instance, Hoare's monitors [13] can be easily implemented and used in the Sather lock statement although they are not directly supported by the programming language.

Java is similar to Sather in many respects, but the Sather lock statement is more general and more powerful than Java's synchronized blocks and methods. A detailed comparison of the Sather lock statement with synchronization constructs of other languages can be found in [14].

3 The LMS Architecture

The major design decisions for the LMS involve selecting the representation (either active or passive) and choosing granularity.

3.1 Active versus Passive

A lock management system can be either active or passive. Functions of an active LMS are performed by one or more designated threads. A passive LMS consists of a set of (passive) objects which are accessed by threads requesting locking operations. A passive LMS might satisfy many non-conflicting requests concurrently. However, synchronization is necessary to serialize critical sections of conflicting requests. For an active LMS, less coordination is necessary since synchronization objects may be permanently assigned to managing threads performing LMS functions. An active LMS approach may lead to a simpler overall design, but has negative performance implications. The disadvantage of an active LMS is that for each lock statement two context switches are necessary (from the requesting thread to an LMS thread and then back to the requesting thread) which is generally more expensive than providing safety for passive lock management.

The Sather runtime presented in this paper uses a passive LMS. A passive LMS was chosen because thread context switch time on the supported platforms is much longer than the time required to synchronize threads by low level synchronization primitives.

3.2 Granularity

Sather is a parallel programming languages designed to run on a network of symmetric multiprocessors, called clusters. This adds another dimension to the synchronization problem, namely granularity. Levels of granularity for a distributed LMS can be: (a) one instance per system; (b) one instance per cluster; (c) one instance per lock or group of locks; (d) one instance per lock per cluster. An earlier implementation of the LMS used a single instance per system [15]. But since even a single instance per cluster (b) leads to serialization for threads that otherwise would run completely asynchronously, an LMS of fine granularity is a better choice for fine-grain parallel systems.

The finest granularity distributed synchronization algorithms (d) for mutual exclusion were extensively investigated in earlier studies (see overview in [17]). However, since in the general case lock statements deal with local synchronization objects, an LMS with slightly lower granularity (c) was chosen. If needed, the finest granularity can still be achieved on the synchronization policy level.

3.3 Internal Structure

The details of the LMS design are best characterized by the interfaces of its subsystems. In addition to the lock interface ($LOCK), two more interfaces are introduced: a *lock manager interface* ($LOCK_MANAGER) and a *management access* interface ($MANAGEMENT_ACCESS). An object implementing the lock manager interface is uniquely assigned to each lock. It is the only object that may use the lock interface and call methods on the synchronization object. A lock manager may manage more than one lock and a lock manager for a particular synchronization object may change dynamically at runtime. In order to ensure thread safety and atomicity of locking operations, threads executing in the lock manager object must be protected by mutual exclusion.

When a lock is created, it is assigned a new lock manager. When the lock is used in a statement with more than one lock for the first time, a new manager is created. This new lock manager takes over all locks used in the statement.

The lock manager interface contains methods for acquiring, trying and releasing locks:

```
abstract class $LOCK_MANAGER is
    acquire(locks:ARRAY2{$LOCK}):INT;
    release(locks:ARRAY2{$LOCK});
    ...
end;
```

A two-dimensional array of locks is passed to these methods containing a row for each branch of the Sather lock statement. An empty row is used for an else branch. If there exists no such row that all its locks could be acquired, the executing thread is blocked until this condition changes. The condition is re-evaluated on each call to acquire or release. Since all changes of the synchronization object state are made known to the associated lock manager, this design is free of race conditions. The actual implementation of thread blocking uses the blocking mutual exclusion primitive available on a particular hardware platform.

The lock manager interface provides other methods for safe transition between lock managers when synchronization objects are used in combinations. The way a lock manager keeps sleeping threads is not specified because different locking policies might require different representations. However, the lock manager transition interface is strictly enforced.

Since lock managers are generated dynamically, the compiler cannot place them in static memory. Consequently, the locks themselves have to serve as entry points for lock management at runtime. For this purpose, all locks have to implement another in-

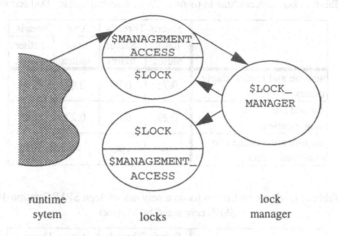

runtime		lock
sytem	locks	manager

Figure 1: Access graph of the lock management

terface, the management access interface. It provides a safe way for finding a current manager for a given synchronization object.

When entering a lock statement, the runtime system accesses the appropriate lock manager via one of the statement's locks (Figure 1).

This design makes both simple and common cases fast. The standard Sather library locks implement not only the $LOCK and $MANAGEMENT_ACCESS interfaces, but also the $LOCK_MANAGER interface. Unless they are used in combinations in complex lock statements, simple locks are managed by specialized lock managers. This saves indirections and dynamic dispatches necessary in the general, but less common case. When locks are used in a more complex context, a more general lock manager is created and takes over the locks.

4 Implementation

The Sather parallel runtime system is highly portable. It includes its own user-level thread package, called Active Threads. Active Threads are available for Sparc V8 and V9, Intel 386 and higher, DEC Alpha AXP, and HPPA 1.1. Alternatively, proprietary thread systems, such as Solaris Threads and POSIX threads [16] may be used as a compilation target. Tables 1 and 2 show execution times of some basic Sather synchronization operations. For comparison, the times for the blocking mutual exclusion operations of the underlying synchronization primitives are also presented.

The numbers in Table 1 were obtained on a Sparc Ultra Enterprise 5000 server with 8 processors. The overhead for a Sather lock is relatively small, in some cases Sather locks are even faster than the basic primitives. The reason for this is that Sather lock managers are protected by spinlocks [18] which are much faster than blocking mutex primitives. If all locks can be acquired, the lock manager does not need to use mutex primitives; these are only required, if a thread has to be blocked. Since acquiring locks

Table 1: Lock access time in μs on an 8cpu Sun Enterprise 5000 server.

	Solaris Threads		Active Threads	
	basic mutex	Sather mutex	basic mutex	Sather mutex
acquire and release single mutex	0.73	0.99	0.83	0.66
try and fail single mutex	0.29	0.82	0.27	0.49
acquire and release two conjunctive locks	—	1.43	—	0.76

Table 2: Lock access time in μs on a network of 4cpu SPARCstation-10 SMPs connected by Myrinet

	Solaris Threads		Active Threads	
	basic mutex	Sather mutex	basic mutex	Sather mutex
acquire and release single local mutex	1.5	2.13	2.0	1.33
try and fail single local mutex	0.6	1.90	0.50	1.03
acquire and release two conjunctive locks locally	—	3.29	—	1.68
acquire and release single remote lock	—	309	—	221
try and fail single remote lock	—	158	—	112

in combinations is supported only by Sather, the corresponding cells for underlying thread systems are left blank.

Table 2 shows the same measurements on a distributed platform of SPARCstation-10 symmetric multiprocessors connected via Myrinet, a low latency packet switched network. Numbers for access to remote Sather locks are also provided.

5   Conclusion

The Sather lock management system displays a clean, object-oriented design that supports sophisticated synchronization policies while preserving high efficiency. The system is fully implemented and runs on several parallel and distributed platforms.

References

1. W.E. Snaman Jr., D.W. Thiel: The VAX/VMS Distributed Lock Manager. Digital Technical Journal, September 1987.
2. A. Bhide, S. Shepler: A highly available lock manager for HA-NFS. Proc. Summer 1992 Usenix Conf., Usenix Association, Berkeley, pp. 177-184, 1992.
3. L. Daynes, O. Gruber, P. Valduriez: Locking in OODBMS clients supporting nested transactions. Proc. 11th Intl. Conf. on Data Engineering, IEEE Computer Society Press, Los Alamitos, pp. 316-323, 1995.
4. M. Aldred, I. Gertner, S. McKellar: A Distributed Lock Manager on fault-tolerant MPP. Proc. 28th Hawaii Intl. Conf. on System Sciences, Vol. 1, IEEE Computer Society Press, Los Alamitos, pp. 134-136, 1995.
5. T. Stiemerling, T. Wilkinson, A. Saulsbury: Implementing DVSM on the TOPSY Multicomputer. Symposium on Experiences with Distributed and Multiprocessor Systems, Newport Beach, 1992, Usenix Assoc., Berkeley, pp. 263-278, 1992.
6. G. Hermannsson, L. Wittie: Fast Locks in Distributed Shared Memory Systems. Proceedings of the 27th Hawaii Intl. Conf. on System Sciences, Vol. 1, IEEE Comp. Soc. Press, Los Alamitos, pp. 574-583, 1994.
7. G.D. Parrington, S.K. Shrivastava, S.M. Wheater: The design and implementation of Arjuna. Computing Systems 8(3):255-308, 1995.
8. B.N. Bershad, E.D. Lazovska, and H.M. Henry: Presto: A System for Object-Oriented Parallel Programming. Software-Practice and Experience 18(8):713-732,1988.
9. S. Omohundro: The sather programming language. Dr. Dobb's Journal 18(11):42-48, 1993.
10. C. Szyperski. S. Omohundro, S. Murer: Engineering a programming language: The type and class system of Sather. J. Gutknecht (Ed.), Programming Languages and System Architectures, Springer-Verlag, pp. 208-227, 1993.
11. C.A.R. Hoare: Towards a theory of parallel programming. Operating Systems Techniques, C.A.R. Hoare, R.H. Perrot (Eds.), Academic Press, London, pp. 61-71, 1972.
12. J. Gosling, B. Joy, G.L. Steele: The Java Language Specification. Addison-Wesley, Reading, Mass., 1996.
13. C.A.R. Hoare: Monitors: an operating system structuring concept. Comm. of the ACM 17(11):549-557, 1974.
14. C.-C. Lim: A Parallel, Object-Oriented System for Realizing Reusable and Efficient Data Abstractions. Ph.D. thesis, University of California at Berkeley, 1993.
15. C. Fleiner: Parallel Optimizations: Advanced Constructs and Compiler Optimizations for a Parallel, Object-Oriented, Shared Memory Language Running on a Distributed System. Ph.D. thesis, University of Fribourg, Switzerland, 1997.
16. The Institute of Electrical and Electronics Engineers. Portable Operating System Interface (POSIX) - Part 1: Amendment 2: Threads Extensions [C Language]. POSIX P1003.4a/D7. April, 1993.
17. T. Johnson: A Performance Comparison of Fast Distributed Mutual Exclusion Algorithms. Proceedings of the 9th International Parallel Processing Symposium, IEEE Computer Society Press, Los Alamitos, California, 1995.
18. T.E. Anderson: The performance of spin lock alternatives for shared multiprocessors. IEEE Transaction on Parallel and Distributed Systems 1(1):6-16, January 1990.

Experiences with an Object-Oriented Parallel Language: The CORRELATE project

Wouter Joosen, Bert Robben, Henk Van Wulpen, and Pierre Verbaeten

K.U.Leuven, Belgium

In this paper, we introduce CORRELATE, a programming language that is being used in the development of various distributed memory applications, ranging from parallel search to numerical simulations.

In principle, CORRELATE exploits the know how of concurrent object-oriented programming languages with high level synchronisation primitives. This is an essential element to support concurrency between the computational entities.

In the development of parallel object-oriented applications, CORRELATE has been used to program both application objects and system objects. The integration of both aspects in one programming environment is enabled by designing a meta-level architecture which models the interaction between application entities and the execution environment. Our paper introduces CORRELATE and will discuss code examples from different applications, as well as from system objects that have been developed in the language framework.

1 Introduction

Object-oriented methodologies are becoming the mainstream approach for software development in general because they enable software houses to develop applications in a man power effective way. This applies all the way down from analysis and design to implementation, testing and maintenance. (The latter includes the evolution that is caused by ever changing requirements.) The increased modularity that is inherent to the use of objects is a crucial advantage when modifying specific parts of the product, when porting applications with respect to varying infrastructure, when isolating errors, etc.

Concurrent object-oriented programming is a very promising area in computer science, because of its expressive power and its natural approach to model the real world. This property should simplify the development of various kinds of applications, for instance in the area of computer simulations. An additional advantage is related to the mapping of application objects on a distributed memory (HPC) machine because concurrency is intrinsically present in the object model.

In this paper, we illustrate this approach and the way HPCN application programs are developed using CORRELATE, which is a concurrent object-oriented language that supports applications running on distributed memory systems. We show CORRELATE programs that implement parallel applications. We discuss the architecture of the underlying object support system and the way it enables the development of optimal system software for a specific application.

Section two summarises the main features of CORRELATE. Section three illustrates the language with key elements of an MD simulator. Then we will discuss an application that solves optimisation problems using cooperative search agents. Section four explains the key characteristics of the run time system. We conclude in section five.

2   CORRELATE in a Nutshell

Applications are developed in two phases. First, the application developer must describe his application guided by a *computational model*. The result is a description corresponding to the problem domain of the application. In this phase, hardware architecture aspects (like distribution) are completely hidden for the application programmer, who models a set of *application objects*.

Secondly, an optimal *execution environment* must be described to target the application to the specific architecture it is running on. The objects that will be added are *system objects*, which are available as reusable components. Load managers are an example of such system objects.

This section summaries the computational view of CORRELATE. In a first phase of application development we are concerned about the logical decomposition of the application in autonomous objects that interact with each other. These autonomous objects reflect the natural abstractions of the application's problem domain. In this section, we will briefly sketch how CORRELATE supports concurrent object-oriented technology. CORRELATE is a class-based concurrent object-oriented language. The CORRELATE syntax is influenced by C++: our first prototype was built as a preprocessor for C++. Today, our prototypes also support the CORRELATE language extensions for Java. We will only discuss language annotations as far as they are specifically related to concurrency and distribution.

```
active Shopper{
...
interface: // Reactive Behaviour
        void ShowTicket();
behaviour:
        bool IsTicketBought();
        for ShowTicket() precondition IsTicketBought();
implementation:
        ...
};
```

Fig. 1. The Shopper class

A CORRELATE class mainly is a code template that enables the instantiation of individual agents[1]. In the computational view, the focus is on location independent interaction in a global agent space.

In its basic form, a CORRELATE agent is pretty much comparable with a concurrent object. An important concern in concurrent environments is the synchronisation of objects. Depending on the state of the object, a specific operation may or may not be executed. In other words, in a certain state an object can accept only a subset of its entire set of operations in order to maintain its internal integrity. This issue can be expressed with synchronisation constraints that reflect application specific semantics of the object. The description of synchronisation constraints inherently enforces the programmer to reveal state information of an active object. The problem of specifying synchronisation constraints therefore is related to the inherent conflict between encapsulation (one of the basic features of OO) and concurrency.

Our basic approach in designing the CORRELATE language is to define a view on an active object that reveals more state information than the amount that would be required in a sequential model, while maintaining encapsulation as much as possible. CORRELATE objects therefore expose an intermediate abstraction layer which basically corresponds to an abstract state machine.

At the level of a class definition, the language's syntax provides a behaviour section that describes the abstract states. Figure 1 describes a Shopper agent that buys airplane tickets for its owner.

In principle, each abstract state is represented by a boolean selection operation that can determine whether an object is in the corresponding state or not. For the shopper, *IsTicketBought* determines whether the agent has actually bought the ticket. The application programmer then implements the mapping of the actual implementation (the private data members) on the state machine. A precondition can be specified for each constrained operation. This precondition can use the abstract state and the parameters of the operation. When a precondition is just a boolean expression that uses abstract states, it can be inserted in the behaviour section of the class header. (This has been illustrated in Figure 1 for the *ShowTicket* operation: the agent can only show the ticket when it has actually bought it.) Otherwise it is coded as any other operation of the class. Operations without a precondition are unconstrained.

A main feature of a CORRELATE agent is its capability to perform so-called autonomous operations. Autonomous operations reflect the autonomy of the instantiated agents. The operational semantics of an autonomous operation causes its invocation each time it is finished. Figure 2 extends the shopper agent with autonomous operations; for instance *BuyTicket*, which models the task that is delegated to the shopper agent.

Finally, we briefly discuss object interaction. *Object interaction* happens by sending messages (invoking operations). Different models for message passing exist. Synchronous message passing blocks the sender's activity until the op-

---

[1] The code fragments that will be shown through this paper are based on CORRELATE classes, because they essentially model large collections of similar agents.

```
active Shopper{
autonomous: // Autonomous behaviour
     void BuyTicket() { ticket = shop $ Buy(destination); }
     void GetInfo();
     void EvaluateInfo();
interface: // Reactive Behaviour
     void ShowTicket();
behaviour:
     bool IsTicketBought();
     bool IsGoodAgencyKnown();
     for Buyticket() precondition IsGoodAgencyKnown() && NOT IsTicketBought();
     for GetInfo() precondition IsEnoughInfoGathered();
     for ShowTicket() precondition IsTicketBought();
implementation:
     ...
};
```

Fig. 2. The Shopper agent

eration is executed to completion in the receiver object. With asynchronous message passing, the sender does not have to wait for the completion of the invocation, thus increasing concurrency. CORRELATE supports both synchronous ($-operator) and asynchronous (
@-operator) message passing. The application programmer must explicitly specify what kind of message passing (s)he prefers.

3 An Example: MD Simulation

This section illustrates sample code from an MD simulator. Before discussing the corresponding CORRELATE class, we briefly characterise the application domain.

The fundamental aim of molecular dynamics is to predict the behaviour of condensed matter from a knowledge of the forces acting between the component atoms. This is done by numerically solving the Newtonian equations of motion to obtain the time evolution of the molecular trajectories.

A challenging molecular dynamics application is the simulation of protein molecules (for example in the context of drug design). A protein molecule can be thought of as a complex flexible mechanical system subject to a number of forces between parts of itself and the environment. For each type of force, specific force calculation algorithms have been developed by application domain experts. In this paper, we will only focus on short-ranged forces as this is sufficient to illustrate the difference between the two MD simulators we discuss in the paper.

The category of short ranged forces contains short ranged repulsions between atoms which is often modelled using the Lennard-Jones potential. The forces between two atoms is a function of the distance between them. The longer the distance the less important the force will be in the force field. Thus, applications use a cutoff range in which atoms interact. A particular atom will only consider neighbouring atoms which reside within this range.

The code fragment below shows the *SimulationBox* class which represents a collection of particles. The aim of the *TimeStep* operation is to compute the forces on all of the encapsulated particles, to determine the new positions and velocities of these particles and to communicate a subset of the new positions to *neighbouring* objects[2]. The latter part involves invoking *DeliverPositions* (reactive behaviour) on these objects. The synchronisation constraints are as follows: each time *TimeStep* is finished, the object enters a WAITING state until all the needed positions have been delivered. Then the object becomes READY and a new *TimeStep* can be executed.

```
active        SimulationBox {
autonomous ://Autonomous Behaviour
     TimeStep() {
          ComputeForces();
          ComputeVelocitiesAndNewPositions();
          CommunicateNewPositions();
          //Deliver new position to all simulationBoxes within a range
     }
interface : //Reactive Behaviour
     void DeliverPositions(ParticleList);
behaviour:
     bool IsREADY();
     bool IsWAITING();
     for TimeStep() precondition IsREADY();
     for Deliver()    precondition IsWAITING();
implementation:
     ParticleList _myparticles;
     ChargeList _mycharges;
     MoleculeList _mymolecules;
...
}
```

Fig. 3. The SimulationBox class in a coarse-grained decomposition

---

[2] These are the objects that need these values for their own force calculations.

4 Example 2: Optimisation Using Multiple Search Agents

A second example demonstrates the use of multiple concurrent and cooperating objects (agents) for solving combinatorial optimisation problems such as scheduling and transportation.

The problem can be described as searching a solution space for a configuration that maximises (or minimises) a given objective function. Even though exact algorithms exist, they are all but unusable for real world problems due to excessive execution times. More promising are so-called meta-heuristics such as simulated annealing and genetic algorithms which cannot guarantee optimality but generate near-optimal solutions in acceptable time.

We designed a framework which merges these meta-heuristics by encapsulating them in cooperative agents. While no single meta-heuristic works well for every possible problem instance, performing concurrent runs with different agents will not only improve the final solutions, but also allows comparison and adaptation of these agents in order to speed up the search.

Figure 4 shows a generalized CORRELATE class definition for these agents. A search agent basically fetches partially optimized solutions from a shared pool (*GetConfiguration*), improves them using its proper algorithm (*OptimizeStep*) and puts the improvements back in the pool for further processing by other agents (*PutConfiguration*). A lower priority meta-search runs in parallel, analyzing the performance of the search agents (*GetStatistics*) and adapting them if apropriate (*Adapt*) [3].

5 The Run Time System

Real world distributed memory applications are complex pieces of software. One of the major problems is due to the explicit notion of distribution in the software architecture: apart from base level objects that model the problem domain of an application, new entities have to be programmed to deal with inherent problems that are caused by distribution. For instance, application-specific performance optimisations and fault tolerance support are often required. In the CORRELATE project, we have modelled the run time system of an application as a meta-level architecture.

Programs that manipulate (control) other programs (say applications) are often termed meta-programs. A clear example of a meta-program definitely is a language run time system. In OOP, metaobjects constitute a meta-program. The objects that are manipulated by metaobjects are often called base-level (or application-level) objects. Note that base-level objects can in turn be metaobjects (and vice-versa).

An architecture is called a meta-level architecture when the meta-level objects are explicitly available for inspection (observation) and modification. In

---

[3] This meta-search is not related to the CORRELATE meta-level architecture described in the next chapter. However it is the latter that allows control of the scheduling of the meta-search.

```
active SearchAgent{
autonomous: // Autonomous Behaviour
      void GetConfiguration() { current = pool$Get(..); }
      void OptimizeStep();
      void PutConfiguration() { poolPut(best); }
interface: // Reactive Behaviour
      void GetCurrentBest();
      Stats GetStatistics();
      void Adapt();
behaviour:
      bool IsSearching();
      bool IsImproved();
      bool IsAdapting();
      for GetConfiguration() precondition !IsSearching() && !IsAdapting();
      for OptimizeStep() precondition !IsAdapting();
      for PutConfiguration() precondition IsImproved();
      for Adapt() precondition !IsSearching();
implementation:
      Configuration current, best;
};
```

Fig. 4. The optimisation search agent

other words, a meta-level architecture enables the modification of the meta-system, possibly during the execution of the application. To support a meta-level architecture in an object-oriented design, one can define a metaobject protocol (hereafter MOP) that defines the set of metaobjects and their interactions (protocols). Specialised support systems can be built by specialising the predefined metaobjects.

In our project, we have experienced the power of MOPs in the development of adaptable and extensible support software for distributed memory applications. CORRELATE has been used to program system objects that are called by meta-objects and that can implement application-specific behaviour in the run time system. For instance, we have implemented load balancing subsystems that manage the work load distribution in an application.

6   Conclusion

The CORRELATE language has been used in the prototyping of multiple HPC applications, ranging from typical scientific simulations to solvers for optimisation problems using cooperative search agents. Moreover, the language has been used in the development of specific subsystems to optimise the execution of a particular application.

In the CORRELATE project, the role of the meta-level architecture is crucial as it enables the seamless integration of application objects and system objects.

This is a key issue in supporting the development of both application and system objects in a single and comprehensive environment.

Towards a Parallel C++ Programming Language Based on Commodity Object-Oriented Technologies

S. Matsuoka[1], A. Nikami[2], H. Ogawa[2], and Y. Ishikawa[3]

[1] Tokyo Institute of Technology
[2] University of Tokyo
[3] RWCP

Abstract. To become widespread, parallel computing should be based on advanced commodity technology, and parallel languages are no exception. But parallel languages must also support a wide range of parallel programming styles, ease of programming, and high performance. We show that C++ can support a wide range of parallel programming styles without special language extensions. More concretely, we used MPC++, a parallel dialect of C++ extended using only templates and inheritance, to create a prototype class/template library which supports three kinds of parallel programming styles. We tested the library performance with representative programs in each of the programming styles.

1 Introduction

Parallel processing in the past required special-purpose hardware, and custom-designed programming languages, compilers, and operating systems. Both hardware and software were expensive and difficult to learn about, maintain, and use. Further, due to rapid increases in the performance and the quality of commodity computer technologies, special-purpose parallel computers such as MPPs lag behind parallel workstation and PC clusters, which have better price-performance, reliability, and ease of use.

We believe that the same commodity trend will hold for the software infrastructure of parallel programming. However, in order for commodity technology to be useful, it must be applicable to a wide range of problems, be easy to use, allowing a smooth transition from existing the software base, and maintain high performance relative to existing special-purpose software. Only if commodity software fails to satisfy these requirements should special-purpose software be built.

Our objective is how learn how far we can push the commodity envelope for high-performance parallel computing. Specifically, we are building prototype class and template libraries for MTTL-based MPC++ v.2.0[1, 2], which support a wide range of parallel programming styles. In this paper, we report on the usability and performance of the libraries. Benchmarks on the RWC workstation cluster, and the RWC PC cluster (both using a Myrinet network), showed performance almost comparable to that using customized compilers on the AP1000+, a system with processors similar to those in the workstation cluster. Our findings show a high usability for commodity languages, compilers, and operating systems. Customization of the existing software base can be minimized, allowing high-performance parallel computing to be much more general and ubiquitous.

2 Assumptions on Software "Commodity-Ness"

To satisfy "commodity-ness" requirements, we establish these assumptions on the language, the compiler, and the operating system. We use: standard programming languages, and no preprocessors; a general-purpose (sequential) C++ compiler; only standard, portable operating system features; and commodity-based hardware. We compare results against custom languages and compilers, perhaps running on special-purpose hardware.

We use MPC++ v.2.0, which extends C++ only via a template library called MTTL. Based on MPC++, we also use classes, inheritance, templates, operator overloading, and macros (all supported by standard C++) to further extend capabilities. Compared to HPC++, [3] whose Level 0 language also uses template libraries for language extensions, we aim to support a wider range of parallel programming styles, such as MPI-based message passing, Split-C like asynchronous (loose) SPMD, traditional concurrent object-oriented programming, and distributed shared-memory (DSM) programming. We also plan to support other parallel programming styles, and to integrate our compositional traverser framework[4] for data distribution. In this paper we describe results for Split-C emulation and DSM, neither of which are supported by HPC++.[1]

Asynchronous Loose SPMD—Split-C: In contrast to MPI, Asynchronous Loose SPMD (AL-SPMD) does not block the sender, allowing extensive overlap of communication and computation. Remote memory access also allows overlap via remote asynchronous put() and get() operations. Split-C[5] supports the AL-SPMD style of programming. It extends C with several constructs such as *split-phase assignments, global pointers,* and *spread arrays.* In the original Split-C, the split-phase assignments are supported by primitive operations of the underlying Active Messages layer: put (acknowledged remote write), get (remote read), and store (non-acknowledged remote write) operations. The sync operation performs global barrier synchronization of put and get operations. In our prototype library, we fully emulate the Split-C semantics using the MPC++ language primitives on commodity cluster hardware, and show that we achieve performance comparable to the Split-C implementation on AP1000+.

---

[1] We have also implemented an MPI version using the low-level communication facilities of MPC++[9]. Benchmarks have shown its performance to be superior to TCP/IP-based MPI implementations. The latest version as of this writing on the PC cluster achieves well over 100MBytes/sec throughput when using the zero-copy capability of the underlying PM communication driver. Detailed results will be reported in another paper.

Cache-coherent Distributed Shared Memory: CC-DSM is a popular programming model for parallel computing. Recent advances in weak coherency models, such as Lazy Release Consistency[8], have made implementation of CC-DSM practical without special-purpose memory consistency hardware. In our prototype library, we achieve object-wise coherency semantics instead page-based semantics usual in procedural languages. There has been other work on object-based DSM systems ([7], for example), in our work we seek the best portable performance achievable with low-latency, commodity communication platforms. Our library currently supports only strict consistency semantics with write-invalidation; lazy coherency semantics [6] is planned.

3 Implementation of Prototype Class/Template Libraries

MPC++ provides primitive constructs for parallel programming with objects and threads such as global pointers, thread management, reduction, and barriers. For a distributed-memory environment, it supports remote memory read _mpcRemoteMemRead(), remote write _mpcRemoteMemWrite(), and synchronizing structures Sync. All the constructs are provided as templates and class libraries with no compiler modifications of the underlying C++.

Using MPC++/MTTL and the extension capabilities of C++, we designed and implemented prototype class/template libraries to support different styles of parallel programming.

3.1 Asynchronous Loose SPMD(Split-C++)

We have implemented a Split-C emulation template library, called Split-C++, supporting global pointers, split-phase assignments, and spread arrays. The Split-C++ global pointer is implemented by extending the MPC++ global pointer. For split-phase assignments, Class mpcSplitC implements the following methods:

Split-C	Split-C++
put	mpcPut(pe,raddr,laddr,size)
get	mpcGet(pe,raddr,laddr,size)
sync	mpcSync()
store	mpcStore(pe,raddr,laddr,size)
store_sync	mpcStoreSync()

Split-phase assignment is performed through the mpcSplitC object, one of which is assigned to each processor. By using simple macros we hide the object, allowing us to have a straightforward one-to-one translation between Split-C code and Split-C++ code:

```
// macros to make things easier
#define put(sc,r,l) \
    sc.mpcPut(r.getPe(),r.getLaddr(),&l,sizeof(l));
```

```
#define sync(sc)    sc.mpcSync();
mpcSplitC msc;
foo() {
  GlobalPtr<int> rm;
  int lc;
                     // On Split-C:
  put(msc,rm,lc); //   rm := lc;
  sync(msc);      //   sync();
}
```

Spread arrays in Split-C are also emulated. In a Split-C spread array definition: "double A[10]::[10][10];", value in brackets to the left of :: gives the array distribution, and the right side is the array definition. Arrays are distributed cyclically. For example, among eight processors, processor 0 will hold elements A[0][0][0] through A[0][9][9] as well as elements A[8][0][0] through A[8][9][9]. In Split-C++, the DistArray class template implements split arrays. DistArray allocates the array with the Init method, and supports standard array accesses via overloading [ ]. Thus the Split-C array definition above is simply translated to:

```
DistArray<double> A; A.Init(1,2,10,10,10);
```

3.2 Cache-Coherent DSM

We support cache-coherent DSM programming with a generic interface for classes Cache and *Shared-Object*, form which all cache-coherent, sharable objects inherit. In particular, class SharedDistArray implements transparent access to cache-coherent arrays of different software cache line sizes.

Our DSM support is currently incomplete. For this paper we implemented a prototype to measure achievable performance of our platforms. We have found that strict coherency semantics hampers performance, so we plan to re-design the library so that more efficient coherency protocols (such as lazy-release) become plug-ins. The prototype assigns a 128-entry, fully set-associative cache to each processor. Cache line size is 256 bytes (32 elements for arrays of doubles).

The Cache class implements the following interface for write-invalidate cache:

isCached(pe,laddr,size): Check if remote data is cached.
readRemote(pe,raddr,laddr,size): If remote data is cached, read locally; else, do remote read into cache, notify owner.
writeRemote(pe,raddr,laddr,size): Do remote write, ask owner to invalidate copies.
writeLocal(dest,src,size): Do local write. If data had copies, ask owner to invalidate them.
clearCache(): Clear the cache.

A new class CachedDistArray has the same interface as DistArray, which does not support caching.

4 Evaluation

Several benchmarks were used to investigate whether our commodity-based library could compete with special-purpose software and hardware. The environments were (1) the RWC Sun workstation cluster with 36 SPARCstation 20/71 (75MHz,32MB) nodes interconnected with Myrinet (LANai4.0), and (2) the RWC PC cluster with 64 Pentium (166Mhz) nodes, also interconnected with Myrinet(LANai4.1). The base performance of the RWC-developed low-level PM messaging layer for PC clusters exceeds 100MB/s under zero-copy situations.

The benchmarks were MM (matrix multiply) and CANNON (from the Split-C example suite), and 1-D FFT (from the SPLASH-2 suite). For MM (the original is referred to as SMALL-MM), we investigated the scalability of the WS and PC clusters by increasing problem size to 2048×2048 matrices (called LARGE-MM).

Modifications to the original Split-C and Splash-2 code to translate to our syntax was small. For MM, fewer than 40 of 700 lines were changed; for CANNON, 30 of 500 lines were changes; for FFT, 90 lines out of 1500. The translations were usually one-to-one and syntactically straightforward.

4.1 Split-C++ Benchmarks

Figures 1, 2 and 3 are the results of the Split-C++ benchmarks for SMALL-MM, LARGE-MM, and CANNON, respectively. Each shows total execution time (communication and computation).[2]

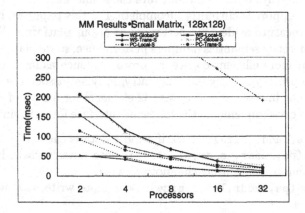

Fig. 1. SMALL-MM in Split-C++ (128,128)

---

[2] We were unable to run CANNON on the PC cluster.

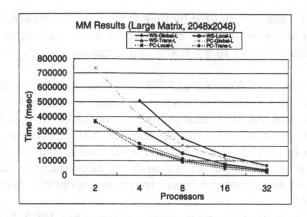

Fig. 2. LARGE-MM in Split-C++ (2048,2048)

SMALL-MM: This achieves good speedup on both the WS and PC clusters, and is significantly faster compared to MPI MM performance. While the performance of the WS and PC clusters are remarkably similar, careful observation reveals that the PC cluster has better node performance, and the WS cluster is better at communication.[3]

LARGE-MM: Both the WS and PC clusters scale well. In fact, local computation shows almost linear speedup. Communication does not quite scale as well, however, primarily due to global synchronization overhead. Comparing the WS and PC cluster, we observe that the PC cluster shows better node performance, while communication performance remains almost identical.[4]

CANNON: Speedup degrades because communication increases faster than node time decreases, due to smaller problem sizes on each node. In comparison, performance of Split-C on the Fujitsu AP1000+[10] is more resilient to decrease in problem size due to better communication performance of the AP1000+ for fine-grain data transfer.[5] We hope that future commodity network products will eliminate this performance difference.

4.2 FFT Benchmark with DSM objects

Figure 4 shows the result of Splash-2 1-D FFT benchmark with DSM objects. We compare three versions of the program: *WS-bulk*, where communication schedul-

---

[3] We do not yet know why the latter is so, since low-level benchmarks show that the PC cluster has better communication performance, both in terms of latency and throughput.

[4] This is also for unknown reasons, since the PC cluster has better low-level communication performance for large data transfers.

[5] MPC++ on the WS cluster achieves 10MB/s for 4KB blocks and 18MB/s with 20KB blocks. In contrast, the AP1000+ performs at 23MB/s for smaller 1KB blocks.

ing has been manually optimized by bulk transfers; *WS-Nocache*, where all remote accesses are done in 8-byte chunks and are not cached; and *WS-cached*, where the cache line is set to 256 bytes, and is managed by the write-invalidate protocol.

Comparing the *WS-Nocache* and *WS-cache*, we see that caching definitely improves performance. However, when the number of processors increase, lower cache-hit ratio and communication overhead overwhelms the benefit of caching. Also, *WS-bulk* is twice as fast as caching, showing that the overhead of our library is significantly larger than those reported for previous systems[6]. Although overhead may be more apparent in our system due to faster communication hardware and libraries, we should nevertheless strive for better optimization and more efficient protocols.

Figure 4 shows that we get no speedup for either *WS-bulk* or *WS-cache*. This is because communication overhead nullifies the increase in the number of processors.

As is with SMALL-MM, both WS and PC clusters show very similar performances. Observing in more detail, we see that *PC-nocache* is faster than *WS-nocache*, while *WS-cache* on two processors is half as fast as *PC-cache*. While the former could be attributed to smaller latency of the PC cluster communication, the latter phenomenon is still not accounted for, since the bulk transfer version does not show this anomaly.

5   Summary and Future Work

We have investigated the use of commodity object-oriented software technology for high-performance parallel computing. Although our library is still new and not well optimized, it shows comparable performance to special-purpose languages on conventional MPPs. There are still, however, areas for improvement.

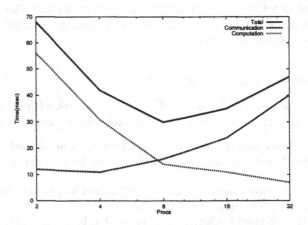

Fig. 3. CANNON in Split-C++ (64,128)×(128,96)

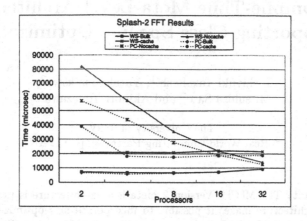

Fig. 4. Complex FFT (1024)

For example, for split arrays, we lose performance due to the lack of inlining the [] operators. Still, we believe that the performance gap will narrow as commodity compiler technology and our library design evolves. Only then we should seek optimizations relevant only to parallel programming.

References

1. Y. Ishikawa et. al.: "MPC++," In Gregory V. Wilson and Paul Lu, editors, *Parallel Programming Using C++*, pp. 427–466. MIT Press, 1996.
2. Y. Ishikawa: "Multiple Threads Template Library—MPC++ Version 2.0 Level 0 Document," Technical Report TR96012, RWC, September 1996, www.rwcp.or.jp/people/mpslab/mpc++/mpc++.html.
3. D. Gannon et. al.: High Performance C++, www.extreme.indiana.edu/hpc++.
4. N. Sato, S. Matsuoka, J.M. Jézéquel, and A. Yonezawa: "A Methodology for Specifying Data Distribution using only Standard Object-Oriented Features," *Proc. ACM/IEEE Int. Conf. on Supercomputing*, Vienna, Austria, pp. 116-123, 1997.
5. D. E. Culler et. al.: "Parallel Programming in Split-C", *Proc. Supercomputing '93*, Nov. 1993.
6. Daniel J. Scales et.al.: "Shasta: A Low Overhead, Software-Only Approach for Supporting Fine-Grain Shared Memory," *Proc. 7th Int. Conf. on Architectural Support for Programming Languages and Operating Systems*, October, 1996.
7. Paul Lu: "Aurora: Scoped Behaviour for Per-Context Optimized Distributed Data Sharing," *Proc. 11th th Int. Parallel Processing Symp. (IPPS)*, Geneva, Switzerland, pp. 467-473, 1997.
8. P. Keleher, A. L. Cox, and W. Zwanepoel: "Lazy Release Consistency for Software Distributed Shared Memory," *J. Parallel and Distributed Computing*, Vol.29-2, pp. 126–141.
9. F. B. O'Carroll: MPIXX User's Guide, www.rwcp.or.jp/people/ocarroll/usingmpixx.html.
10. Kobayashi et.al.: Implementation and Performance Evaluation of Split-C on AP1000+, *2nd Hokke2 '95 Workshop*, IPSJ SIGHPC Notes, 1995 (*in Japanese*).

A Compile-Time Meta-Level Architecture Supporting Class Specific Optimization

Toshiyuki Takahashi[1], Yutaka Ishikawa[2],
Mitsuhisa Sato[2] and Akinori Yonezawa[1]

[1] The University of Tokyo
[2] Real World Computing Partnership, JAPAN

Abstract. The MPC++ Version 2 meta-level architecture is proposed. The architecture makes it possible to incorporate new optimizers into a compiler. A library designer can provide an optimizer specific to his class/template library in the library header file. A library user may use such a high-performance library by including the header file. A meta-level programming library, called DMSF, is designed to facilitate programming optimizers. As an example, a distributed array class in a parallel environment is introduced and its optimization program is presented. The optimizer is simple to describe, and it is effective. Evaluation using the CG kernel of NAS Parallel Benchmark shows that the program optimized by the meta-level optimizer runs almost as fast as the equivalent C program.

1 Introduction

To support high-performance, parallel object-oriented programming, researchers have attempted C++ language extensions in two ways: (1) syntax extensions, such as pC++[12], CC++[3], and ICC++[5], and (2) introduction of compiler directives (or pragmas) such as the HPC++[1] level 0 specification. However, it is difficult to design and implement parallel description abstractions appropriate to describe all parallel application fields.

ABC++[10], HPC++Lib, and MPC++ [9] are projects that use the C++ class/template feature. They provide parallel description primitives without language extensions. POOMA[11] and KeLP[6] provide parallel application-oriented libraries. They have the benefit of object-oriented programming such as encapsulation, inheritance, and polymorphism, but have the disadvantage of low performance. Our experiments[7] show that a C++ program is 1.2 to two times slower than the same application written in C. One difficulty for a compiler optimizer is difficulty in analyzing side effects of object operations.

We have designed and implemented an extendable C++ programming language, MPC++ [8, 9], that has a compile-time meta-level architecture. The MPC++ has been redesigned to support building optimizers, so that a class/template library designer can build an optimizer specific to his library and supply the optimizer as a plug-in module to the compiler. The plug-in module is supplied by the library header file. the optimizer analyzes and optimizes the program code using the MPC++ meta-level facility. Thus a library designer may design a parallel library along with its specific optimization facilities.

In this paper, we propose the new MPC++ meta-level architecture, called MPC++ Version 2 meta-level architecture. In Sec. 2, an overview of the MPC++ Version 2 meta-level architecture is presented. Section 3 presents an example of class-specific optimization for a `DistArray` class that implements a distributed container in a parallel environment. In Sec. 4, a meta-level programming library called DMSF is designed to facilitate programming optimizers. An optimizer for `DistArray` is programmed using the library. To show the power of the optimizer, the CG Kernel of the NAS parallel Benchmark has been implemented using the `DistArray` class library. Section 5 shows that the program optimized by the meta-level optimizer runs almost as fast as an equivalent C program.

2 An Overview of MPC++ Meta-level Architecture

The MPC++ Version 2 level 1 specification defines a meta-level architecture[1] that consists of an abstract compiler described in C++, its modification facility, and a facility to traverse syntax trees. The interface to these facilities is called the MPC++ meta-object protocol. The user may write a meta-level program using the meta-object protocol. The program is a plug-in module in the sense that it is incorporated into the MPC++ compiler at compilation time.

Figure 1 shows an example of a new optimizer written as a plug-in module. The keyword $meta tells the MPC++ compiler to shift to the meta-level. An optimizer is defined as a class derived from the `Traverser` class in meta-level. The MPC++ compiler invokes the `traverse` method, whose argument is a syntax tree, defined in the `Traverser` class. The `traverse` method traverses a syntax tree and executes a method for each node if the method is defined in a subclass of `Traverser`. The method for a syntax tree node may replace the syntax tree node. In this example, the `MyForStmtOpt` class defines the `forStmtVisit` method, which is invoked when the object finds a `for` statement node. The `forStmtVisit` method receives a `ForStmt` syntax tree object as the first argument, transforms it, and returns the result using the second argument.

When the compiler parses $meta { ... } in Fig. 1, it interprets the statements in braces. In this example, an object of `MyForStmtOpt` is created and then the `regTraverser` meta-level primitive function is invoked. Then the compiler invokes the `traverse` method of the object each time the compiler parses a declaration or a function in the source program.

3 An Example of Class Specific Optimization

To show an optimization program using the MPC++ meta-level architecture, an array object in a distributed memory-based parallel machine is introduced. An

[1] MPC++ Version 2 provides two levels, 0 and 1. Level 0 specifies parallel description primitives implemented with the C++ template feature.

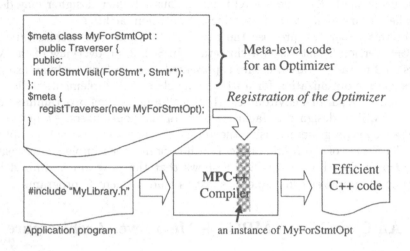

Fig. 1. Incorporation of a new optimizer.

array is partitioned among processors and represented by the `DistArray` object. Each processor has the `DistArray` object with local storage for the partitioned area. If access is to local storage, the object returns a pointer to the storage. If access is for the remote storage, the remote storage reference system is invoked.

As shown in Fig. 2, the `DistArray` object contains member objects `data`, `lower`, `upper`, and `dist`. The `DistObj` pointed to by `dist` defines the partitioning scheme among processors. The `data` member keeps the pointer to the partial array storage. The members `lower` and `upper` represent an index boundary of the partial array storage. All member variables are initialized by the `init` method and are not changed by other methods.

The `operator[]` method of `DistArray` implements transparent access to elements. The method returns an object of class `GlobalPtr<double>` provided in the MPC++ level 0 library[9]. An assignment to a `GlobalPtr` object expression is replaced with the remote memory write primitive. Reference to a `GlobalPtr` expression is replaced with the remote memory read primitive.

Figures 2-2 and 2-3 show an example using `DistArray`, and its optimized code. The optimized code is efficient because it does not generate `GlobalPtr` operation calls when local elements in `DistArray` objects are accessed, and loop invariants are moved out from the loop body.

To generate the optimized code, an optimizer must know the following propositions:

```
class DistArray {
public:
    double      *data;
    int         lower, upper;
    DistObj     *dist;
    void        init(DistObj*);
    //
    GlobalPtr<double> operator[](int);
    int         getLower() { return lower; }
    int         getUpper() { return upper; }
};
```

2-1. DistArray Class

```
Block       block(10000);
DistArray   x, y;

x.init(&block);
y.init(&block);
for (i = x.getLower(); i < x.getUpper(); i++) {
    x[i] = y[i];
}
```

2-2. Original Code

```
Block       block(10000);
DistArray   x, y;

x.init(&block);
y.init(&block);
int         _xl = x.getLower();
int         _xu = x.getUpper();
double      *_xd = x.data;
double      *_yd = y.data;
if (x.dist == y.dist ) {
    for (i = _xl; i < _xu; i++) {
        _xd[i - _xl] = _yd[i - _xl];
    }
} else {
    for (i = _xl; i < _xu; i++) {
        _xd[i - _xl] = y[i];
    }
}
```

2-3. Optimized Code

Fig. 2. DistArray class and its example code.

- When x is an object of DistArray and i is the loop variable bounded by x.getLower() and x.getUpper(), x[i] means access to local storage. x[i] can be replaced with x.data[i - x.lower].
- When x and y are objects of DistArray and i is the loop variable bounded by x.getLower() and x.getUpper(), y[i] means access to a local storage if x.dist == y.dist. y[i] can be replaced with y.data[i - x.lower] if x.dist == y.dist.
- Member variables, lower, upper, dist and data are invariant in a loop if the loop does not include any invocation of the init method.

No C++ compiler can find out these propositions automatically because it cannot analyze data flow among objects which include invocations of separately compiled code. In this example, the implementation of GlobalPtr includes system calls that implement communication among processors, so a compiler cannot analyze data flow among DistArray objects.

On the other hand, the DistArray class designer knows these propositions, and can program an optimizer based on them using the MPC++ meta-level architecture.

4 Implementation of the Optimizer

To facilitate writing optimizer code using the meta-level architecture, we are designing and implementing class libraries for meta-level programming. In this section, the DMSF library, consisting of classes DMSF_Analyzer and DMSF_Translator, is introduced to support transformations of method invocations in a for loop. Then an overview of an optimizer specific to DistArray using the library is presented.

An object's method call is invariant in a loop if the loop contains no expression to affect the object. In the case of DistArray objects, getLower and getUpper method calls are loop invariant if the loop does not contain the init method. DMSF_Analyzer is responsible for marking such invariant method calls in loops. A derived class of DMSF_Analyzer must define method isSuppressMethod that tells which method invocation affects an object. Such a method affecting an object is called a *suppress method*. The derived class may define the method isTrustedMethod to tell which method invocations do not affect an object.

After analyzing a syntax tree with an object derived from DMSF_Analyzer, the tree is passed to an object derived from DMSF_Translator to transform expressions. DMSF_Translator has the mostOuterLoop member variable which keeps the outermost loop among those whose bodies do not include any *suppress method* invocation. The user may program funcExprVisit, whose first argument is a method or function call expression syntax tree, to hoist a method call to before a loop pointed to by mostOuterLoop, if the call is marked as an invariant method call.

Figure 3 shows an overview of an optimizer specific to DistArray. The optimizer is built by the designer of DistArray using the DMSF meta-level programming library and includes propositions described in the previous section. Analyzer_DistArray has methods isSuppressMethod and isTrustedMethod. Method forStmtVisit is defined to recognize a loop index and its boundaries. A variable carrying the loop index will be stored in LOOPVAR attribute of the for loop. A variable carrying the object that defines loop boundaries will be stored in BOUNDARY attribute of the for loop. The method funcExprVisit of Translator_DistArray hoists a method call to getLower or getUpper in DistArray from a loop pointed to by mostOuterLoop if the call is marked as an invariant call by the DMSF_Analyzer. To expand the operator[] method, LOOPVAR and BOUNDARY attributes are used.

When attributes LOOPVAR and LOOPOBJ of a for loop are i and p respectively, expression p[i] in the loop means local storage access. The optimizer will translate the expression to p.data[i - p.lower]. Expression q[i] in the loop means local storage access if q takes the same distribution pattern as p. The optimizer will translate the expression to (p.dist == q.dist, p.data[i - q.lower], q[i]). The triplet is a conditional expression that includes loop-invariant condition. To expand the triplet we supply the Specializer class, also derived from the Traverser class. Specializer uses the triplet to switch between two version of the loop, one using the local reference and the other using the non-local reference.

Using the three traversers Analyzer_DistArray, Translator_DistArray and Specializer, we generate efficient code for uses of the DistArray class. The last part of Fig. 3 shows the meta-code to register three traversers into the MPC++ compiler.

```
$meta class Analyzer_DistArray : public DMSF_Analyzer {
  bool isSuppressMethod(FuncExpr* fx) {
    // If fx calls DistArray::init(DistObj*), it returns true;
    // otherwise it returns false.
  }
  bool isTrustedMethod(FuncExpr* fx) {
    // If fx calls DistArray::operator[](int), it returns true;
    // otherwise it returns false.
  }
  int forStmtVisit(ForStmt* fs, Stmt** ret) {
    // If fs matches with the following pattern,
    // for( lv = Obj.getLower(); lv < Obj.getUpper(); lv++)
    //    { /* lv is invariant. */ }
    // set lv and Obj on attribute LOOPVAR and BOUNDARY of fs respectively.
  }
};

$meta class Translator_DistArray : public DMSF_Translator {
  int funcExprVisit(FuncExpr* fx, Expr** ret) {
    // Code that expands the method invocations
    // DistArray::operator[](int), DistArray::getLower()
    // and DistArray::getUpper().
    // using LOOPVAR, BOUNDARY attributes and mostOuterLoop.
  }
};

$meta {
  registTraverser(new Analyzer_DistArray);
  registTraverser(new Translator_DistArray);
  registTraverser(new Specializer);
}
```

Fig. 3. An overview of an optimizer specific to DistArray.

5 Performance

To show the effectiveness of the optimizer specific to the DistArray class library, we implemented the CG kernel of the NAS Parallel Benchmark Version 1 (class A) on the RWC workstation cluster consisting of 36 Sun SS20's (75MHz) connected via Myrinet. GNU G++ 2.7.2 is used as the backend compiler of MPC++.

Table 1 shows the results. Both "Not Optimized" and "Optimized" rows give times (and ratios to times for a C version) of MPC++ programs using the DistArray class without and with its class-specific optimizer, respectively. The "C Program" row gives times for a C version of the GC kernel. The program using the DistArray class without class-specific optimization runs about 1.8 times slower than the same application in C. Our class-specific optimizer reduces the overhead of the DistArray class and achieves 5% to 9% slower execution than the program written in C.

Table 1. Execution times for the CG kernel.

	16PE	32PE
Not Optimized	27.630 (1.72)	15.663 (1.77)
Optimized	16.886 (1.05)	9.609 (1.09)
C Program	16.030 (1.00)	8.840 (1.00)

6 Related Work

Sage++[2] has a feature that constructs an abstract syntax tree from C++ source code. It does not provide a way to change the syntax of the input language dynamically. As far as we know, it provides no high-level mechanisms for operating on the syntax tree, such as pattern matching or code analysis.

OpenC++[4] has a meta-level architecture to change the semantics of method invocation. To support a meta-level programmer, it provides a meta-level programming library named Meta Object Protocol. However, the library does not have a way to analyze the context of the method-changing semantics. Moreover, it is a bootstrapping compiler in the sense that a new compiler must be built using the meta-level architecture. Thus, several new compilers will be built if several class-specific optimizations are implemented.

7 Summary

In this paper, the MPC++ Version 2 meta-level architecture has been proposed, and a DMSF meta-level programming library to facilitate building class-specific optimizers have been designed. The meta-level architecture makes it possible to incorporate a plug-in module of a class-specific optimizer to the compiler. The DistArray class and its class-specific optimizer were introduced as an example. The optimizer is simple to describe and it is effective. Results using the CG kernel of NAS Parallel Benchmark show that code written using the DistArray with its optimizer runs 5% to 9% slower than the code written in C, while the code using the DistArray without the optimizer runs about 1.8 times slower than the C code.

We are currently designing and implementing a meta-level programming library for data flow analysis that uses class-specific information such as dependency of methods. The library supports the designers of class-specific optimizers to make more powerful optimizers.

Acknowledgements We sincerely thank Rod Oldehoeft for his helpful comments on drafts of this paper. We also thank Shigeru Chiba, Kenjiro Taura, Hidehiko Masuhara, Tatsuro Sekiguchi, and Tomio Kamada for their helpful advice on this work.

References

1. P. Beckman, D. Gannon and E. Johnson. Portable Parallel Programming in HPC++. *This article is obtained via http://www.extreme.indiana.edu/hpc++.*
2. F. Bodin, P. Beckman, D. Gannon, J. Golwals, S. Narayana, S. Srinivas and B. Winnicka. Sage++: An Object Oriented Toolkit and Class Library for Building Fortran and C++ Restructuring Tools. In *Proceedings of OONSKI'94*, 1994.
3. K. M. Chandy and C. Kesselman. CC++: A Declarative Concurrent Object-Oriented Programming Notation. In *Research Directions in Concurrent Object Oriented Programming*. MIT press, 1993.
4. S. Chiba. A Study of a Compile-time Metaobject Protocol. PhD thesis, Graduate School of Science, The University of Tokyo, Japan, 1996.
5. A. Chien, U. Reddy, J. Plevyak and J. Dolby. ICC++ – A C++ Dialect for High Performance Parallel Computing. In *Object Technologies for Advanced Software*, vol. 1049 of *LNCS*, pp. 76–95. Springer-Verlag, 1996.
6. S. J. Fink, S. B. Baden and S. R. Kohn. Flexible Communication Mechanisms for Dynamic Structured Applications. In *Proceedings of IRREGULAR'96*, 1996.
7. J. Gerlach, M. Sato and Y. Ishikawa. Using the C++ Standard Template Library for High Performance Applications. *IPSJ SIG Notes*, 97(37), pp. 19–24, 1997.
8. Y. Ishikawa, et. al. Design and Implementation of Metalevel Architecture in C++ – MPC++ Approach –. In *Proceedings of Reflection'96*, pp. 141–154, 1996.
9. Y. Ishikawa, Multi Thread Template Library – MPC++ Version 2.0 Level 0 Document –. Real World Computing Partnership, TR–6012, 1996. *This technical report is obtained via http://www.rwcp.or.jp/lab/pdslab/mpc++/mpc++.html.*
10. W. G. O'Farrell, F. C. Eigler, S. D. Pullara and G. V. Wilson. ABC++. In *Parallel Programming Using C++*. MIT press, pp. 1–42, 1996.
11. J. V. W. Reynders, P. J. Hinker, J. C. Cummings, S. R. Atlas, S. Banerjee, W. F. Humphrey, S. R. Karmesin, K. Keahey, M. Srikant and M. Tholburn. POOMA. In *Parallel Programming Using C++*. MIT press, pp. 547–588, 1996.
12. S. X. Yang, D. Gannon, P. Beckman, J. Gotwals and N. Sundaresan. pC++. In *Parallel Programming Using C++*. MIT press, pp. 507–545, 1996.

An Object-Oriented Approach to the Implementation of a High-Level Data Parallel Language[*]

Matthias Besch, Hua Bi, Gerd Heber, Matthias Kessler, Matthias Wilhelmi

Parallel and Distributed Systems GMD Laboratory Berlin, Germany
GMD FIRST, Rudower Chaussee 5, D-12489 Berlin
{mb,bi,heber,mk,wilhelmi}@first.gmd.de

*This research is supported by the *Real World Computing Partnership*

Abstract: This paper presents an object-oriented approach to a high-level programming language for parallel scientific computing and its realization by the compilation and runtime system PROMOTER. At its language level, PROMOTER supports a relaxed data parallel, aggregate object model, extending the usual features of object-orientation to its new concepts for specifying and operating on aggregate or distributed objects and data transfer (communication) in between. PROMOTER's runtime system is realized as a generic class library, which efficiently supports different degrees of symmetry in application structures and dependence patterns. Both runtime system and (coordination) language are based on an imperative, statically typed, object-oriented language (C++).

1 Introduction

A programming model for scientific parallel computing that claims to be high-level needs, above all, problem-adequate expression tools for modelling the spatial (or algebraic) structures that occur in an application. A characteristic of data parallel languages is that they offer some form of grouping mechanism in order to apply lifted operations on collections of data. Most prominent representatives are vectors and arrays which, however, often do not reflect the application structures properly. And concepts like indirect indexing can not be considered as high-level, since they ignore any structural aspects.

PROMOTER introduces a new concept called *index space*, which extends the indexable structure of an array to an arbitrarily shaped subset of some Z^n. Like conventional arrays, they permit a compact representation avoiding the disadvantages of pointers. Unlike arrays or (nested) vectors, they are far more flexible and *directly* support also non-rectangular and irregular structures, while preserving their geometric or algebraic structures. At the same time they can be uniformly used to model both data aggregates and dependences (communications) in between.

Based on the notion of index spaces, PROMOTER offers a number of high-level programming concepts, which are briefly discussed in the following section. The focus of the paper, however, lies on the (object-oriented) design principles of their implementation in the compiler and runtime system (Section 3). Section 4 shows that such a translation of high-level language constructs into executable code can be done efficiently. Finally, a few remarks on related work are made.

2 PROMOTER Programming Language Concepts

PROMOTER [1][8] is a coordination language embedded in C++ (or other object-oriented language). The key feature is the *topology* construct whose intention is two-fold. Generally, a topology defines a so-called *index space*, an arbitrarily shaped subset of some multi-dimensional, cartesian integral space. In particular, a topology is intended for modelling an application's spatial data structures as well as the dependence relations within or between data domains.

Its first application can be considered as a collection mechanism grouping data on which replicated function operate as a whole. A topology may be parameterized and consists of a part specifying its dimension and range followed by a constraints part cutting out the intended subset.

```
topology Tridiag (int N) : 0:N , 0:N {
  $i, $j I: abs(i-j) <= 1; // select the set { (i, j) I Ii-jI <= 1 }
};
```

In analogy to inheritance and subtyping, topology declarations may be reused by means of *composition* and *specialization*, which means that the index space of a topology may be composed of the cartesian product of one or multiple already defined topologies being further restricted by additional constraints.

```
topology EverySecondTridiag (int N) : Tridiag (N) {
  $i, $j I: i%2 == 0; // selects every second element of Tridiag
};
```

Dependent on the degree of symmetry inherent to an application's data structures, other construction mechanisms can be used. So, at its one end, index spaces can be constructed by enumerating their individual elements. Once declared, a topology can be used (multiple times) to define so-called *aggregate types* and *aggregate objects*, for example

```
Class Element { ... } ; // includes a method f (args)
EverySecondTridiag(1000)<Element> obj;
```

An aggregate object is subject to a partitioning and mapping process and is thus potentially distributed. An operation on it (or some selected subset thereof), for example obj.f(args), triggers a data parallel operation being independent on its topology points. Exploiting the *polymorphism* properties of the host language (element subtyping, virtual functions,...) is one way (at the language level) to relax the simple data parallel execution model.

Communication is realized by so-called data transfer expressions as part of a replicated function's argument evaluation. Since data dependence can be modeled as some subset of the cartesian product between source and destination index space, it is itself an index space and can thus be handled by the same language construct *topology*.

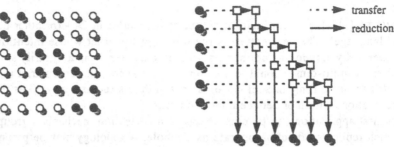

Fig. 1. Tridiag(5) as data and communication topology

Note that, frequently, data dependencies exist between finitely many *communication neighbours* (for example in a grid), which can be ideally modeled by some constraint expression of a potentially *infinite* topology.

```
topology ShiftX(int x): Grid, Grid {
    $a, $b, a+x, b; // $a and $b are formal indices
};
```

3   Implementation Framework

The programming environment is implemented in two major building blocks: the PROMOTER compiler (PROMC) and the PROMOTER Runtime System (PRTS) which consist of the PROMOTER Runtime Library (PRL) and the PROMOTER Abstract Machine (PAM). PROMC works as a source-to-source translator and generates C++ code containing execution primitives of the PRL. The PRL is implemented on top of the PAM which provides portable communication and synchronization primitives on different platforms. We focus here on the runtime model of the PRL.

Our first implementation is based on the SPMD model with a lock-step synchronization for architectures with distributed memory and message passing. Note that the PROMOTER language itself does not determine the execution model.

3.1   Object Model

A PROMOTER program represents a global view onto *aggregate objects*. In the runtime system, *aggregate objects* will be represented by *descriptors* and *carriers* in an object-oriented way. *Descriptors* contain information about *index spaces* and their partitions in the SPMD environment, while *carriers* contain local data elements of *aggregate objects*.

Descriptors are defined by the orthogonal trinity of Topology, Mapping and Distribution by means of functionality decomposition.

The Topology class captures the spatial structure of *index spaces*, e.g. a *valid* method should determine which indices belong to an *index space*, and a global iterator should allow iteration over all indices in an *index space*. A particular topology can be expressed as a subclass of the embedding topology. Depending on the complexity of the index space the valid method reduces to the evaluation of a boolean expression (e.g. for a rectangular grid) or incorporates a search algorithm in irregular cases. A topology object can change its shape at runtime implementing appropriate methods for adding and removing points.

A PROMOTER program is split into tasks for execution in a SPMD environment, where the set of elements which can be accessed locally by a task is called *domain*. The Mapping class assigns a *domain* number to an index. The splitting of the global view into domains is controlled by a mapper. For different levels of complexity and dynamicity different techniques can be applied, for static applications the mapping can be performed at compile time. Depending on Topology and Mapping, the Distribution class captures partition of the index space according to the mapping. It provides transformations of global indices into local, linearized coordinates which guide storage of local data elements for an efficient organization of a task's local elements. It also provides a local iterator over all indices assigned to the own domain. The Mapping object may change during runtime.

The *carrier* of an *aggregate* object is defined by the class template of Disobj (distributed object) which is parametrized over the Distribution and the element type. (There is a further refinement of this view depending on whether the element type is static or dynamic.) The runtime view of *aggregate* objects can be illustrated by the following figure:

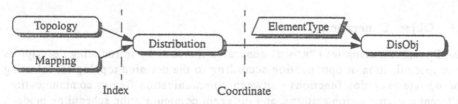

Fig. 2. Object Model

In this runtime model, we support object sharing (one Distribution object may be shared by many Disobj objects, Topology objects and Mapping objects may be shared by Distribution objects), polymorphism (the implementation of Disobj class is generic on the interface of Distribution), and specialization (descriptors can be specialized in a problem-oriented way by a compiler, and then used without modifying the implementation code of Disobj class and the operations).

Since *descriptors* are *mobile* objects preserved over the whole runtime, the runtime model supports dynamic topology and mapping, as well as dynamic load balancing. For example, an index can be added or removed from a Topology object, a Mapping object can be generated at compile time, and then changed by a *mapper* at runtime, and a Distribution object can also be changed at runtime.

3.2   Operations on Distributed Objects

In the PRL, operations on Disobj objects are provided by function templates. They can be classified as *communication-free* operations and *collective communication*. The collective communication functions are divided into different communication patterns such as one-to-one, one-to-many (expansion), and many-to-one (reduction). Collective communication functions are implemented in two phases. In the first phase, communication scheduling generates send and receive queues of local coordinates according to two Distribution objects and a communication topology to specify the communication pattern for their designated Disobj objects. In a second phase, the communication operation realizes overlapping communication and computation.

A default communication scheduling is implemented also in parallel with two operation modes: sender-initiated or receiver-initiated. Sender or receiver-initiated communication scheduling collects the local coordinates at sender or receiver locally (therefore it is a scalable) and then transfers the queues to the corresponding receiver or sender. The communication scheduling can also be optimized and then used in the PRL without modifications in the function templates.

Because we decouple the partitioning information from distributed objects, our communication scheduling is independent of distributed objects. In this way, the result of communication scheduling can be reused by the collective communication operations based on the same distributions and the same communication topology. In a lot of data-parallel applications, the reuse of communication scheduling will largely improve the efficiency, especially when lifting a communication scheduling out of a loop.

3.3   Object Generation

The PROMOTER compiler (PROMC) generates the code by introducing descriptors with specialization or optimization according to the declared topologies, selecting appropriate operation functions (between communication-free or communication, different communication patterns, and different communication scheduling modes), and optimizing the code, e.g. by reuse of communication scheduling. Because our runtime model is constructed in the object-oriented way, it provides polymorphism for an easy implementation and specialization for an efficient execution of data-parallel applications.

4 Results

We have implemented the first version of the PROMOTER environment. The system has been ported to our in-house testbed MANNA, to the IBM SP/2, to the CRAY T3E and to SUN workstation cluster using either ethernet (MPICH) or the Myrinet hardware (MPC++, MTTL [5]) for communication. Several optimizations concerning the mapping of aggregate objects, data parallel operations and communication scheduling were tested and implemented.

Several applications (CFD, FEM, div. solvers) have been written in PROMOTER. The first experiences proved a good performance and scalability for different problem classes.

time / s		8 nodes	16 nodes	24 nodes	32 nodes
	PROMOTER	21.63	12.14	7.99	7.47
SP/2	PROMOTER (opt.)	9.46	5.46	3.45	3.38
	reference impl.	N/A	5.60	3.48	2.34
	PROMOTER	8.6	4.7	3.6	
T3E	PROMOTER (opt.)	4.5	2.6	1.7	
	reference impl.	6.5	2.6	N/A	

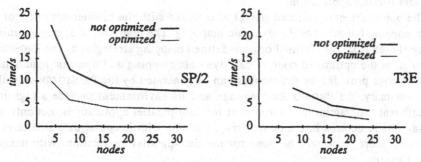

Fig. 3. Results for the NASPAR CG-Kernel Class A

We implemented the conjugate gradient benchmark from the NASPAR benchmark suite [7] in PROMOTER. The kernel employs the multiplication of a sparse matrix with a vector. It is a typical example of unstructured grid computation with irregular communication. The performance results for the reference implementation, the PROMOTER program, and an optimized PROMOTER version (communication scheduling external to loop) are shown in Figure 3.

5   Related Work and Conclusion

The PROMOTER environment is related to a number of approaches on programming distributed memory parallel computers in an object-oriented manner, such as HPC++[4], pC++[6], MPC++[5], IC++[2] and EC++[3].

A PROMOTER program gives a global and abstract view of the problem. It hides all aspects of distribution details. The programmer has only to deal with problem oriented structures. The most distinct construct of the PROMOTER language is the *topology*. It allows a programmer to explicitly model an application's spatial data structures for rectangular and non-rectangular, dense and sparse, or regular and irregular shapes. More than that, the same *topology* construct is also used to explicitly model the dependence relations within or between data domains for an operation. In this way, the definition and use of topologies in programs can provide informations for the compiler and the runtime system, especially for the mapper to efficiently implement PROMOTER programs.

The PROMOTER language is implemented in the object-oriented way. With object-orientation, encapsulation of data distribution and message passing details simplifies its SPMD implementation. Because of generic class templates, efficient implementation can be achieved by specialization of the contained classes for *descriptors* of aggregate objects. For example, different specializations can be applied to implement *topology* or *distribution* for dense arrays and sparse arrays.

The object-oriented runtime model also allows the PROMOTER compiler to generate efficient code by sharing of runtime objects, and by reuse of code. For example, the *descriptors* of aggregate objects can be shared not only to save construction time but also to save memory space, and lifting the communication scheduling out of a loop, if possible, is very important to achieve high performance, especially for irregular data-parallel application.

The object-oriented runtime model also works with the mapper not only for the static mapping but also for the dynamic mapping. The mapping of aggregate objects can be chosen from pre-defined or user-defined mapping strategies by the PROMOTER compiler, or the optimized code to call dynamic mapping and dynamic load balancing primitives provided by the mapper can be generated by the PROMOTER compiler.

In summary, the PROMOTER language and its environment provide a high-level and efficient programming environment for data-parallel applications, not only with regular data structures like a dense array, but also with irregular data structures. In the future, more work will be done for the data-parallel applications with irregular data structures.

References

1. Besch M., Bi H., Enskonatus P., Heber G., Wilhelmi M.: High-Level Data Parallel Programming in PROMOTER, In *Proc. of Second International Workshop on High-level Parallel Programming Models and Supportive Environments HIPS'97*, IEEE-CS Press, Apr 1997.
2. A. A. Chien and J. Dolby, The Illinois Concert System: A Problem-Solving Environment for Irregular Applications, In *Proc. of DAGS'94, The Sym. on Parallel Computation and Problem Solving Environments*, 1994.
3. The EUROPA Working Group on Parallel C++ Architecture SIG, EC++ - EUROPA Parallel C++ Draft Definition, Technical Report 1995.
4. The HPC++ Working Group, HPC++ White Paper, Technical Report TR 95633, Center for Res. on Parallel Computation, Rice University, 1995.
5. Y. Ishikawa, MPC++ Programming Language V1.0 Specification with Commentary - Document Version 0.1, Technical Report TR-94014, Real World Computing Partnership, June 1994.
6. A. Malony, B. Mohr, D. Beckman, D. Gannon, S. Yang, F. Bodin and S. Kesavan, A Parallel C++ Runtime System for Scalable Parallel Systems, In *Proc. of Supercomputing '93*, pages 140-152, IEEE-CS Press, Nov 1993.
7. The NAS Parallel Benchmarks, *http://science.nas.nasa.gov/Software/NPB/*.
8. P. Enskonatus, M. Kessler, Concepts and Formal Description of the PROMOTER Language, Version 2.0, Technical Report TR-96020, GMD-FIRST, March 1997, *http://www.first.gmd.de/promoter/papers/index.html*

A Framework for Parallel Adaptive Finite Element Methods and Its Template Based Implementation in C++

Jens Gerlach, Mitsuhisa Sato, Yutaka Ishikawa
{jens,msato,ishikawa}@trc.rwcp.or.jp

Tsukuba Research Center of the Real World Computing Partnership
Tsukuba Mitsui Building, 1-6-1 Takezono, Tsukuba-shi 305, Japan

Abstract. Finite element meshes are large, richly structured sets whose internal relationships must be visible to different parts of a finite element program. This causes software engineerings problems that increase when adaptive mesh refinement and multilevel preconditioners are applied. Even more problems arise when finite element methods have to be implemented for parallel computers since the meshes have to be mapped onto the hardware topology so that their locality is preserved. We have designed a framework for parallel adaptive finite element methods that centers upon a problem-oriented *index scheme* as a new high level description method for finite element meshes. Within the index scheme, important mesh relations can be expressed by simple algebraic operations in $\mathbf{Z}^n$. We give an overview of the indexing methodology and outline the main parts of the framework. Special emphasis is on the reuse of several C++ template libraries—including standard container classes and the library for data parallel programming of the PROMOTER programming model.

1 Introduction

In this paper, we deal with software engineering problems that are related to the implementation of adaptive finite element methods (FEM) on parallel computers.

Leaving aside pre- and postprocessing, implementing FEM (for scalar boundary value problems) means the treatment of two different, yet closely related problems.

1. The description and manipulation of the finite element meshes I.
2. Performing numerical methods in the vector space $\mathbb{R}^I$.

Adaptive FEM can be characterized by the fact that they repeat these two steps several times—starting with a relatively coarse mesh and performing numerical methods that provide the feedback where the mesh quality has to be increased—until a satisfying solution has been found.

The problem with the finite element meshes is that they are large, richly structured sets. The definition of numerical methods in $\mathbb{R}^I$ that have a reasonable complexity and their efficient implementation crucially depends on an exploitation of the spatial structures of the meshes [12]. Moreover, the structure has to be evaluated to achieve a locality preserving mapping onto parallel architectures. For these reasons FEM (as special adaptive mesh methods) are regarded as difficult to implement on serial and parallel computer architectures [7].

In order to overcome these problems, we have designed a framework for the implementation of parallel adaptive FEM.

Within the framework the application programmer shall be able to concentrate on the particular finite element method ie the used element types, mesh hierarchies and preconditioners. He/she shall not care about the low level aspects of parallel programming—like distributing the data, introduction of ghost points, communication and synchronization. As far as possible this is to be done by lower level software layers. However, the programmer must provide the structural information of the application to them.

Therefore the framework centers upon a problem-oriented *index scheme* as a new high level description method for finite element meshes. We outline the indexing methodology in section 2.

In section 3 we present the main components of the framework. Currently it is used to solve FEM approximations (using linear triangle elements) of the Poisson equation with Dirichlet boundary conditions on an arbitrary polygon by applying iterative methods (conjugate gradient, Richardson). Several preconditioners, including multilevel ones, have been implemented. We report on our experiences with C++ template libraries that have been used for the implementation.

Finally we compare our approach with related work and draw conclusions.

2 A Problem-Oriented Index Scheme for FE Meshes

In this section we present our high level description methodology for finite element meshes. We are speaking of *index schemes* since the essence of our methodology lies in devising problem-oriented embeddings of the components of a finite element triangulation into some $\mathbb{Z}^n$.

Our research has been motivated by the development of the PROMOTER[1] programming model for parallel scientific computing [9]. In PROMOTER the programmer has to use *index spaces* (subsets of $\mathbb{Z}^n$) to model the spatial structures of an application. This has the advantage that both data aggregates and their relations can be expressed by the same language construct (called *topology*).

The proposed index scheme, which was first presented in [1], aims at raising the level of abstraction on which finite element methods are described away from particular implementation techniques towards the problem domain. This

---

[1] PROMOTER is a C++-based data parallel programming model for parallel computers with distributed memory that is aimed at scientific applications. Its development at GMD-FIRST was supported by the Real World Computing Partnership, Japan. PROMOTER makes the logical spatial structures of an application explicit and thus evaluable to the compilation and runtime system, thereby permitting the automatic mapping of them to the physical architecture and the direct generation of a message-passing implementation.

gives the implementor the flexibility and freedom to create high-performance implementations [4].

The representation in this section is twofold. We talk both on the general aspects of the indexing methodology and explain them with the help of triangle elements. To achieve an expressive embedding of the components of finite element meshes into $\mathbb{Z}^n$ we proceed as follows.

First we require that there is a coarse triangulation that represents the combinatorical structure of the domain[2]. Part of this requirement is that all triangulation are contained in the coarse triangulation ie each element of an actual triangulation is a subset of exactly one coarse element. This is guaranteed if the actually used triangulations are created by refinement of the coarse one. This is no severe restriction, since fast iterative solution methods depend on a hierarchy of nested triangulations [12].

Due to the complexity of domain shapes found in real life the coarse triangulation will usually be provided by a computer aided design tool. Since regularity can't be expected for the coarse triangulation it is the best to denote its components by an *enumeration*. This is called the *coarse index scheme* of the triangulation. Figure 1 shows a L-shape domain together with a coarse triangulation.

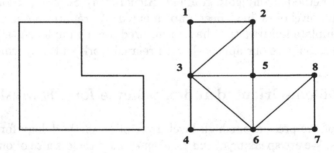

Fig. 1. An L-shape domain and its coarse triangulation. Vertices are indexed by integers (in arbitrary order). Coarse edges and triangles are denoted by ordered pairs (triples respectively) of integers that correspond to their vertices. This kind of coarse index scheme can be applied to other types of elements as well. An advantage of this indexing method is that it is possible to extract the incidence relations ie which faces are shared by neighboring elements.

On each of the elements of the coarse triangulation we choose an indexing technique that is appropriate for the element shape. For example, it will be based on grid coordinates for rectangular elements or scaled barycentric coordinates for simplex elements. We call this indexing method the *local index scheme*. Figure 2

---

[2] Of course, special attention must be given to problems where the combinatoric structure of the application changes during the computation (e.g. crash simulation) [8]. In such cases it might be necessary to create new coarse triangulation during the run of the finite element program.

sketches the foundation of a local scheme for the case of triangle elements. The use of such an *element-oriented* indexing method allows it to express important mesh relations by simple index calculation. This will lead to compact descriptions of finite element meshes and support a symbolic processing of the mesh structure.

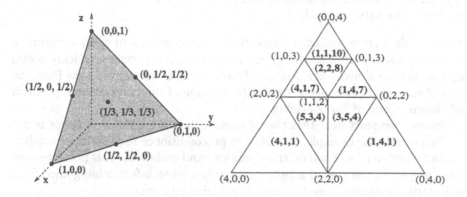

Fig. 2. The local scheme for triangle elements is based on scaled barycentric coordinates. The left picture shows the two-dimensional standard simplex σ^2 as it is embedded into $\mathbb{R}^3$. Introducing barycentric coordinates on a coarse triangle (shown in the right picture) can be interpreted as a mapping onto σ^2. The actual indices of the local scheme are obtained by scaling the barycentric coordinates of a vertex or of the barycenter of a triangle (bold face) by a power of 2. The exponents used reflect the refinement level of the actual triangulation.

The final *global index scheme* that presents the desired embedding into $\mathbb{Z}^n$ is just a combination of the coarse scheme and the local one. Since both the coarse and the local index scheme for triangle elements use triples of integers, the global scheme is realized as an embedding into $\mathbb{Z}^6$. One problem that stems from this combination is that components of the mesh that are shared by several coarse elements have several names within the global scheme. However these ambiguities can be resolved with the help of the incidence relations contained in the coarse scheme. More precisely, this is done by introducing an unambiguous *normal form* (in addition to the *standard form*) for the indices of the global scheme.

3 Design and Implementation of the Framework

The framework will provide a testbed for multilevel preconditioners for iterative methods of large scale elliptic problems that are suited for parallel computer architectures. Corresponding to the main components of a finite element algorithm the framework provides interfaces for

- Domain description
- Representation of the boundary value problem
- Selection of element types
- Representation of the triangulations and node sets
- Grid functions and element matrices
- Discretized differential operator and preconditioners
- General iterative methods

This hierarchy of concepts helps concentrating the complicated mesh structure in a relatively small part of the program. Therefore other parts can quickly access this information if necessary. Another benefit of this hierarchy is that from the point of view of the iterative methods the solution of the finite element problem boils down to one of linear algebra.

However we must note that the relations between the different levels is not one-directional. For example, a multilevel preconditioner requires the introduction and maintenance of several triangulations and node sets. This becomes even more complex in the case of adaptive refinement when information gained from the iterative methods is used to create new triangulations.

3.1 Benefits of the Index Scheme

Within the index scheme the triangulations and node sets turn into subsets I of $\mathbb{Z}^n$. Thus the scheme provides a uniform background for the representation of a wide range of spatial structures.

Using a particular type of finite element corresponds to a set of *simple algebraic relations* that is used to determine the node set that is connected with a given triangulation and this element type.

Grid functions and elements matrices are considered as real- or matrix-valued functions of the node set and triangulation, respectively.

The discretized differential operators correspond to sparse matrices in the vector space $\mathbb{R}^I$. The sparsity pattern depends on the element type and the actual triangulation. They are evaluated using the above mentioned algebraic relations between triangulation and node set caused by the used element type.

3.2 Experiences with a Sequential Prototype

In a prototypical sequential implementation we represented the index sets by the set container of the Standard Template Library (STL) [13]. When using the STL set container with a class K, a function object for comparing K-objects has to be provided. The ordering that results from this comparison object is used both for identifying and quickly accessing the elements from set<K> [13]. When comparing two indexes they are transformed into their normal form, see section 2. By this the programmer can use the more convenient standard form without being bothered by its ambiguities.

We have defined vector and sparse matrix classes on top of the index sets to represent concepts like mesh functions (elements of $\mathbb{R}^I$) or discrete operators (matrices in $\mathbb{R}^I$, e.g. the mass matrix and preconditioners). Elements of these vectors and matrices can be accessed through the indexes of our scheme. We have implemented these classes with the help of the map container (a generic associative array) of the STL. As an example we show the code fragment for the assemblage of the mass matrix from the element matrices. The extremely flexible memory management provided by the map container allows an implementation in which the element matrices have just to be added to get the mass matrix.

```
void SolverTop::assemble_element_matrices()
{
  ElementMatrix::const_iterator i;
  for(i = em.begin(); i != em.end(); i++)  A += (*i).second;
}
```

Many iterative methods for the solution of sparse linear problems depend only on general matrix and vector operation and don't need direct access to their elements. This is exploited in the design of general solver libraries like IML++ which make no assumptions about the data format of the vector and matrix classes [5]. After overloading the basic numerical operators for our vector and matrix classes and providing the preconditioners with a solve method, we were able to use this library of C++ template functions.

We encountered one problem when using the IML++ library, since it uses sometimes vector objects to store temporary *scalar* values (e.g. in the conjugate gradient method). Since the components of our Vector class can only be accessed through indices of our index scheme, we had to modify some internal declarations of the IML++ library. It is not clear to us why the designers of IML++ didn't use the Real template parameter (that is provided by the user anyway) to declare their temporary scalar values.

3.3 Towards a Parallel Implementation

For a first parallel implementation of our framework we will use the runtime library of the PROMOTER programming language. This library provides template classes for distributed objects and template functions for generic operations on them. These functions cover basic data parallel operations, for example, data parallel assignments, data parallel function calls, point-to-point communications and collective communications (reduction, expansion, cross product, and communication product) [9]. Since this data parallel library itself is implemented on top of MPI we can run our program on different platforms without being forced to deal with the low level details of message passing programming.

In the parallel program grid functions, element matrices and the sparse matrices will be distributed over the physical architecture—therefore they are called *distributed objects*. On the library level of the PROMOTER programming model the index space of a distributed object is described by three classes— Topology, Mapping, and Distribution, i.e, the mapping analysis is carried out on the level of index spaces not on the level of distributed objects.

One interesting aspect of the index scheme is that points that are close within in the mesh have a relatively small distance when embedded into $\mathbb{Z}^n$. For this reason we can use special mapping algorithms for example the *topographic* mapping procedure [2].

In a future version we want to use the language layer of the PROMOTER model [10] or of its successor PROMISE[11].

4 Related Work and Conclusions

Chains[3] is a computer language whose primitive types are based on the algebraic topological concepts of cells, cell complexes, and chains. The general aim of the *Chains* project has been to raise the semantic level at which complex systems are programmed and to give a *symbolic* description of a physical problem so that it can be automatically mapped to the computational resources. This and its approach to start with a high level representation of the finite element triangulation is reminiscent of the index scheme approach. However, only direct and simple iterative (Jacobi, SOR) methods are considered as solution methods for the linear system of equations. Modern fast solution methods that are based on adaptive refinement and multilevel preconditioning and their impact on the structure of finite element meshes are not discussed.

Rüde describes in [12] the *fully adaptive multigrid method* (FAMe) for elliptic boundary value problems. He gives an abstract description of multilevel structures and shows how efficient multilevel methods can be developed. His approach of using object-oriented programming techniques aims at raising the level of abstraction, but it introduces new problems. For example, the explicit use of pointers to represent the graph structure of finite element meshes is not appropriate for distributed memory machines. Here, the index scheme is superior, since it does not depend on a particular memory model.

ISIS++ [6], which stands for *Iterative Scalable Implicit Solver (in C++)*, is currently developed at Sandia National Laboratories. It is a portable, object-oriented framework for solving sparse systems of linear equations that originate from large-scale, three-dimensional finite element problems. Its implementation covers the most commonly used Krylov subspace iterative methods with different preconditioners (including multilevel ones). ISIS++ is organized as a class library that includes special version for sequential and parallel platforms. Concerning the representation of finite element methods, ISIS++ defines the concepts of *active*, *shared*, and *external* nodes which directly refer to processors of a parallel computer. This is in contrast to our index scheme approach that aims (in cooperation with a high level parallel programming model like PROMOTER) at removing the burden of such concepts from the application programmer.

Kohn et al describe in [7] a software infrastructure for parallel structured adaptive mesh methods. By "structured" they mean that they utilize a hierarchy of nested mesh levels in which each level consists of many simple, rectangular meshes. They used their application programmer interface (consisting of C++ classes) to the solution of adaptive eigenvalue problems arising in material sciences and analyzed various aspects of the parallel performance of their code.

In contrast to their work, our framework includes the important case of simplex finite elements. Thus it is not restricted to collections of rectangular meshes.

We have pointed out the basic software engineering problems of implementing parallel adaptive FEM. The point is that a good design must center upon a high level description of the finite element meshes. For this we have devised a new problem-oriented index scheme. The prototypical implementation of our framework showed that FEM can be well described with the help of our index scheme. The extensive use of several C++ template libraries simplified the software development drastically.

References

1. J. Gerlach, G. Heber, A. Schramm: *Finite Element Methods in the* PROMOTER *Programming Model*, in Proc. Internat. EUROSIM Conference on HPCN Challenges in Telecomp and Telecom, Delft, Netherlands, June 1996,
2. M. Besch, H.W. Pohl: *Topographic Data Mapping by Balanced Hypersphere Tessellation*, EUROPAR96, Lyon, France, August 1996, LNCS 1124, Springer 1996, RWC-P-95-064,
   http://www.first.gmd.de/promoter/papers/index.html
3. R.S. Palmer: *Chain Models and Finite Element Analysis: An Executable Chains Formulation of Plane Stress*, Technical Report, Cornell University NY, CS TR94-1406, 1994
4. J.M.D. Hill, D.B. Skillicorn: *Lessons Learned from Implementing BSP*, in Proceedings of International Conference of High Performance Computing and Networking (HPCN-97), Vienna, Austria, April 1997
5. J. Dongarra, A. Lumsdainm, R. Pozo, K. Ramington: *IML++ v.1.2 Iterative Methods Library Reference Guide*, available at http://math.nist.gov/iml++/
6. http://www.ca.sandia.gov/isis/isis++.html
7. S.R. Kohn, S.B. Baden: *A Parallel Software Infrastructure for Structured Adaptive Mesh Methods*, Proceedings of Supercomputing'95, San Diego, CA
8. http://www.esi.fr/products/crash/crash.html
9. Besch M., Bi H., Enskonatus P., Heber G., Wilhelmi M.: *High-Level Data Parallel Programming in PROMOTER*, Proc. Second International Workshop on High-level Parallel Programming Models and Supportive Environments HIPS'97, Geneva, April 1, 1997, IEEE-CS Press, RWC-P-96-056,
   http://www.first.gmd.de/promoter/papers/index.html
10. A. Schramm: *Concepts and Formal Description of the PROMOTER Language Version 1.0*, RWCP Technical Report 1994, RWC-TR-94-017,
    http://www.first.gmd.de/promoter/papers/index.html
11. W.K. Giloi, S. Jähnichen, M. Kessler: PROMISE—*High-level Data Parallel Programming of Large Scale Applications*, project proposal, GMD-FIRST Berlin, Germany, 1996
12. U. Rüde: *Mathematical and Computational Techniques for Multilevel Adaptive Methods*, SIAM, 1993
13. Alexander Stepanov, Meng Lee: *The Standard Template Library*, Technical Report, HPL-94-34, April 1994, revised July 7, 1995

Parallel Array Class Implementation Using C++ STL Adaptors

Motohiko Matsuda, Mitsuhisa Sato, and Yutaka Ishikawa
{matu,msato,ishikawa}@trc.rwcp.or.jp

Real World Computing Partnership
Tsukuba Mitsui Building 16F, 1-6-1 Takezono
Tsukuba-shi, Ibaraki 305, Japan

Abstract. STL Adaptors can combine operations and are used in elimination of temporaries in a C++ array class; this technique is known as Expression Templates or Template Closures. Since the technique is dependent on a simple expansion of element references, some difficulties exist in applying the technique to a parallel array class, where distribution with ghost-cells and notation of array sections complicate the expansion of element references. The technique is extended so that it separates element references in two cases to keep the expansion simple in each case. This achieves good performance even with the existence of ghost-cells, whereas the implementation of an existing technique does not support it well because of the required amount of coding. In addition, currying facility of Adaptors is used for supporting nested data structures, where operations are required to nest so that they can be applied to sub-structures. An example shows a mapping of reductions is concisely expressed in a matrix-vector multiplication.

1 Introduction

An array class in C++ is one of the most important classes in the field of high-performance computing. A data-parallel array class such as one proposed in the extended C++ language HPC++ [1] can be implemented as a library without language extensions. Although implementing an array class entirely in a library makes it highly portable, the well-known temporary problem [2] will degrade its performance when arithmetic operators are overloaded.

Recently, good C++ compilers have become commercially available, but temporaries still remain, especially in array or vector classes in which operator implementation includes a loop. For example, we observed that a smart compiler from KAI [3] removes temporaries in all simple cases, but still fails if operator implementation includes a loop.

STL Adaptors [4] allow us to represent a nesting of function applications, and these can be used to delay the evaluation of an expression and to eliminate temporaries. This technique is known as Expression Templates [5, 6], or found in the implementation of the Valarray class [7]. In our previous work, we applied the technique, which we called Template Closure, to eliminate temporaries appearing

in a parallel array class on a distributed memory machine [8]. It is successful for
an array class which is modeled after Thinking Machines Corp's data-parallel C
language C*, where all operations are elementwise and an implementation makes
copies during communications.

In this paper, we present an implementation of a parallel array class with
ghost-cells by employing the extension of the Adaptor technique. Ghost-cells,
also known as guard-cells or envelopes, are the overlapping boundary elements
of a block in a distributed array. They speed up stencil calculations by making
references to the nearest neighbors as local operations. However, the existence
of ghost-cells renders the application of the temporary elimination technique inefficient, because the original technique is dependent on a simple expansion of
elementwise calculations whereas ghost-cells introduce complexity to the expansion. Our extension makes element references simple even with the ghost-cells.

In addition, we present a support for nested data structures through the use
of Adaptors. The nesting of data structures is often used, and sometimes it can
express data-parallel programs very concisely [9, 10]. The currying facility of
Adaptors accommodates the nesting of operations, which is needed in handling
nested data structures. Once an operation is defined as an Adaptor, its application is delayed and it is eligible for combination. This makes the mapping
of mapping operations directly expressible as a combination of Adaptors, which
is different from AVTL [11], another parallel vector class library that supports
nested data structures. In AVTL, it is necessary to explicitly define a new class
of a function object to represent a combination.

In the rest of the paper, Section 2 presents how the temporary elimination
technique is modified to support ghost-cells and array sections, and Section 3
shows comparison of overheads with an existing technique. We show a support
of nested data structures in Section 4, and conclude in Section 5.

2 Implementation of Array Class with Ghost-Cells

We first review the Adaptor technique of temporary elimination using an array
addition as an example, then extend it to support ghost-cells. Figure 1 shows
a tentative sketch of an array class and its overloading of an addition. The
following expression is used as an example showing the expansion of additions,
where a shape_t object denotes the size and distribution of arrays:

```
shape_t *s;
Array<int> X(s), A(s), B(s), C(s);
X = A + B + C;
```

The additions on the right hand side create objects of class expr_add, each of
which represents a sub-expression. The top level object on the right hand side is
referred to as *RHS*. Since expr_add takes operands as template arguments, its
type represents the nesting of expressions. That is, *RHS* has a type including a
type of a sub-expression:

```
expr_add<expr_add<Array<int>, Array<int>>, Array<int>>
```

115

```
/* array class */
template <class T>
class Array {
    typedef T ET;
    T *storage;
    T& elem(int i) { return storage[i]; }

    template <class EXPR>
    operator=(EXPR& rhs) { elementwise assignment }
};

/* object to represent add operation */
template <class EXPR1, class EXPR2>
class expr_add {
    typedef EXPR1::ET ET;
    EXPR1& L;            // left operand
    EXPR2& R;            // right operand
    expr_add(EXPR1& 10, EXPR2& r0) : L(10), R(r0) {}

    /* element reference for addition */
    ET elem(int i) { return L.elem(i) + R.elem(i); }
    void make_copy() { L.make_copy(); R.make_copy(); }
    ET elem_from_copy(int i)
        { return L.elem_from_copy(i) + R.elem_from_copy(i); }
};

/* addition operator overloading */
template <class EXPR1, class EXPR2>
expr_add<EXPR1, EXPR2>
operator+(EXPR1& e1, EXPR2& e2) {
    expr_add<EXPR1, EXPR2> expr (e1, e2);
    return expr;
}
```

Fig. 1. Adaptors for additions and overloading of an addition operator (outline sketch).

Since the assignment of an array is defined as elementwise, the whole expression is equivalent to the following.

```
for ( iterate i over elements )
    X.elem(i) = RHS.elem(i);
```

Since the type of each sub-expression in *RHS* is known at compile time, a compiler can statically expand the access elem(i) by traversing down the nesting of types and generates a code as intended. In this expansion, no temporary is necessary:

```
for ( iterate i over elements )
    X.elem(i) = A.elem(i) + B.elem(i) + C.elem(i);
```

```
/* array section reference */
template <class T>
class expr_indexing {
    typedef T ET;
    Array<T>& A;       // referenced array
    Index index;       // section description
    T *copy;           // copy if needed
    int offset;        // offset calculated

    /* element reference */
    T elem(int i) { return A.elem(i + offset); }
    void make_copy() { make a copy in copy if needed }
    T elem_from_copy(int i) { return copy[i + offset]; }
};
```

Fig. 2. Adaptors for array references (outline sketch).

However, a straightforward application of this technique to a parallel array class would be inefficient. If an array class implementation uses ghost-cells, element reference is sometimes satisfied without a copy, but sometimes needs a copy when element reference spans over distributions. This forces the insertion of a conformance check in an element reference, which makes the implementation unacceptably slow. It is because the technique above depends solely on expanding element references, so that the checks incur a cost in the inner-most loop. Compilers usually cannot move them out of the loop.

In addition, the notation of an array section, a sub-array reference using triplets (start, end, and stride), introduces another check to element references. Array sections may introduce data dependency, which forbids in-place assignments and requires a temporary copy.

The conformance checks necessary in expressions with ghost-cells and array sections are summarized as follows: *compatibility* checks if every reference has the same number of elements, *data dependency* checks if a copy is necessary, and *boundary condition* checks if some reference may span over distributions.

The basic idea in moving these checks out of the inner-most loop is to separate element references into two cases: the first case is for references that do not need a copy and are optimizable, and the second case is for references that need a copy. In both cases, references are expanded statically as in the original technique, but they are selected dynamically.

Figure 1 includes the definition needed to accomplish this separation, and Figure 2 is a class to represent array section references. The class expr_add has two interfaces to access elements. The interface elem() is for cases without a copy, and the other elem_from_copy() is for cases with a copy.

A conformance check is performed at an assignment, and either elem() or elem_from_copy() is selected according to the results of the check. This effec-

```
/* 9-addition of arrays */
X = A0 + A1 + ··· + A8;

/* 9-stencil calculation */
Index I(···), J(···);
X(I,J) = A(I-1,J-1) + A(I-1,J) + ··· + A(I+1,J+1);
```

Fig. 3. Code examples of additions and stencils used in the performance comparison.

tively moves the check out of a loop like a loop invariant motion. The resulting code is shown as follows:

```
if ( conformance check ) {
    for ( iterate i over elements )
        X.elem(i) = RHS.elem(i);
} else {
    RHS.make_copy();
    for ( iterate i over elements )
        X.elem_from_copy(i) = RHS.elem_from_copy(i);
}
```

Precise checks of data dependency are time consuming and not appropriate here, because these checks are performed at run-time. Thus, we reduced the dependency check to guarantee an assignee array does not appear on the right hand side.

3   Comparison of Overheads with An Existing Technique

As an evaluation, we compared the overhead of our implementation with another existing technique. The Aggregate Operators technique used in the A++/P++ library [12] deals with the temporary problem differently. It is a run-time solution and dynamically creates a tree object representing the expression. The run-time routine checks the shape of a created tree and selects an appropriate routine to handle it. It is necessary to prepare the routines for frequently appearing expression patterns in advance. The routines are prepared by the library writer and can be highly optimized.

The comparison uses the code for simple additions and stencil calculations as shown in Figure 3. Table 1 shows the overheads in 9 and 27 additions. In the Aggregate Operators case, the time consumed consists of calls to constructors, element type checks, a hash table look up, and a call to an appropriate routine. The hash table is large enough so that no rehash is required in this comparison. On the other hand, in the Adaptor technique case, the time consumed consists of calls to constructors and a call to the assignment code.

Table 2 compares the time consumed by 9 and 27 additions. The array size is about 1,000 elements for both cases. It should be noted that the Aggregate

Table 1. The overhead of multiple additions. The Adaptor technique has smaller overhead because it is handled by a compiler.

	Adaptor	Aggregate Op.
9-addition	1.14 μsec	25.3 μsec
27-addition	3.30 μsec	175.1 μsec

Table 2. Execution time of multiple additions. The Adaptor technique has comparable or better performance with the Aggregate Operators which use hand-coded routines.

	Adaptor	Aggregate Op.
9-addition	774.5 μsec	759.3 μsec
27-addition	2059 μsec	3364 μsec

Table 3. Execution time of stencil calculations. The Adaptor technique performs better because the Aggregate Operators create temporary copies for stencils.

	Adaptor	Aggregate Op.
9-stencil	1044 μsec	5084 μsec
27-stencil	2383 μsec	13032 μsec

(SPARC Station 20/71 (Solaris 2.5), SPARC Compiler C++ 3.0, option -O)

Operators technique is slower in the 27 addition in spite of the fact that it uses a hand-coded routine. This is due to high register pressure (incurs register spills/refills), because our implementation uses 28 pointers running through each array.

Table 3 shows the time taken in 9 and 27 nearest neighbor calculations using stencils. In the Aggregate Operators, copies are necessary in references to array sections, because the implementation does not include them as candidates for aggregation, which worsens its performance. The Aggregate Operators method itself does not limit the complexity of the expressions handled, but an actual implementation does limit the set of expressions aggregated. Note that we are measuring time under the assumptions of the existing implementation.

4 Nested Data Support with STL Adaptors

By using STL Adaptors to express functions that take function arguments, a curried function can be passed to other operations. Operations on sequences in the STL, such as mappings and reductions, are extended to be combined for supporting nested data structures. For example, a mapping of reductions appears in processing a vector of vectors, which is well expressed as a combination of the operations. However, reductions in the STL are not function objects and cannot be passed to mappings or binders (currying operators), and thus a new class is required to represent the combination.

Figure 4 shows the extended definitions of these operations and their application to sparse matrix vector multiplication [9, 10]. In the last statement of

```
Vector<double>
matvec(CRSMat<double> mat, Vector<double> vec) {
    Vector<double> val = mat.val;
    Vector<int> idx = mat.idx, len = mat.len;
    Vector<double> v = vec(idx);
    Vector<double> p = op_map(op_mul<double>())(val, v);
    Vector< Vector<double> > n = nest<double>(p, len);
    /* mapping reductions on nested data */
    return op_map(op_reduce(op_add<double>(), 0.0))(n);
}


template <class OP>
class op_map {
    typedef Vector<OP::TR> TR;
    typedef Vector<OP::TA0> TA0; typedef Vector<OP::TA0> TA1;
    OP f;
    op_map(OP f0) { f = f0; }
    Vector<OP::TR> operator()(Vector<OP::TA0> v)
        { mapping over a vector v }
    Vector<OP::TR> operator()(Vector<OP::TA0> v0, Vector<OP::TA1> v1)
        { mapping over vectors v0 and v1 }
};


template <class OP, class T>
class op_reduce {
    typedef T TR;
    typedef Vector<T> TA0; typedef Vector<T> TA1;
    OP f; T base;
    op_reduce(OP f0, T& base0) { f = f0; base = base0; }
    T operator()(Vector<T> v)
        { reduction by f on a vector v }
};
```

Fig. 4. Example of mapping of reductions — Matrix vector multiplication in compressed row format (outline sketch).

matvec, the combined function object op_reduce(op_add<double>(),0.0) is created and passed to op_map.

In matrix vector multiplication, the sparse matrix is represented in a compressed row storage (CRS) format: val holds the non-zero elements in the matrix, idx holds the column indices of non-zero elements, and len holds the numbers of non-zero elements in each row. vec is an input vector. The codes are self-evident except for vec(idx) and nest: vec(idx) is a vector indexing to vec which returns a vector of elements vec[idx[i]], and nest takes a flat sequence and nests it by chopping it into given lengths.

Using STL Adaptors, the mapping of reductions can be applied without explicitly defining a new function object, although these definitions of operations are incompatible to the corresponding ones in the STL.

5 Conclusion

We explored a temporary elimination technique for a distributed array class, and compared its performance with the Aggregate Operators technique. The Adaptor technique, while its application is limited, helps compilers eliminate temporaries. The extension to the technique enables to eliminate temporaries even with the existence of ghost-cells. In contrast, whereas the Aggregate Operators technique gives a general solution, it would be beneficial only in specific applications and need burdens of library writers, and thus the implementation does not support ghost-cells well.

In addition, we addressed the support of nested data structures. The STL Adaptors exploit the indirect application of a function by separating the creation of a function object from its application. This is the key to the temporary elimination technique. We found that certain algorithms, such as mappings and reductions, benefit by this indirect application, when they are applied to nested data structures.

References

1. The HPC++ working group. HPC++ Whitepapers and Draft Working Documents. 1995. http://extreme.indiana.edu/hpc++/whitepaper.html
2. M. A. Ellis, and B. Stroustrup. *The Annotated C++ Reference Manual.* (pp.299–303, §12.1c: Temporary Elimination). Addison-Wesley, April 1994.
3. A. D. Robinson. C++ Gets Faster for Scientific Computing. *Computers in Physics,* Vol.10, No.5, 1996.
4. A. Stepanov, and M. Lee. The Standard Template Library. HP Technical Report HPL-94-34, February 1995. http://www.cs.rpi.edu/ musser/stl.html
5. T. Velhuizen. Expression Templates. *C++ Report,* Vol.7, No.26, 1995.
6. S. W. Haney. Beating The Abstraction Penalty in C++ Using Expression Templates. *Computers in Physics,* Vol.10, No.6, 1996.
7. D. Vandevoorde. Valarray<Troy>, 1995. ftp://ftp.cs.rpi.edu/pub/vandevod/Valarry/Documents/valarray.ps
8. M. Matsuda, M. Sato, and Y. Ishikawa. Efficient Implementation of Portable C*-like Data-Parallel Library in C++. *Proc. of the 1997 Advances in Parallel and Distributed Computing,* March 1997.
9. G. E. Blelloch. Programming Parallel Algorithms. *Communications of the ACM,* 39(3), March 1996.
10. G. E.Blelloch, S. Chatterjee, J. C. Hardwick, J. Sipelstein, and M. Zagha. Implementation of a Portable Nested Data-Parallel Language. *Proc. 4th ACM SIGPLAN Symposium on Principles and Practice of Parallel Programming,* May 1993.
11. T. J. Sheffler. A Portable MPI-based Parallel Vector Template Library. RIACS TR-95.04, 1995. ftp://riacs.edu/pub/Excalibur/avtl.html
12. R. Parsons, and D. Quinlan. A++/P++ Array Classes for Architecture Independent Finite Difference Computations. *Proc. of the 2nd Annual Object-Oriented Numerics Conference,* April 1994.

A Multithreaded Java Framework for Solving Linear Elliptic Partial Differential Equations in 3D

Gerald Löffler

Research Institute of Molecular Pathology (I.M.P.)

1 Motivation

Our research in theoretical biophysics [6] recently required us to solve a variant of the Poisson equation, which is a linear elliptic partial differential equation (PDE) in 3D. Ultimately, we want to make our method available as a applet, so we decided to implement a PDE solver in Java.

Another reason for using Java was its promise of easy, portable multithreaded programming, because support for multithreading is built into Java. On a Symmetric Multi-Processor (SMP) this offers the opportunity to painlessly achieve parallel execution, which might be important for the numerically intensive (though certainly not excessive) task of solving a PDE.

The Java framework described here allows the solution of any linear elliptic PDE on a regular cubic domain in 3D. To this end it employs the Full Multigrid (FMG) algorithm [3, 1, 7], which is the most efficient general-purpose algorithm widely used for this problem domain. The framework is easily extensible in every aspect of the PDE algorithm and was designed with interactive Java applets in mind. The framework is implemented in the 1.1 release of Java.

2 The Full Multigrid Algorithm in a Nutshell

This is a brief introduction to the FMG algorithm [3, 1, 7] and the terminology of the FMG framework. The FMG framework deals with solving linear elliptic PDE's on domains in 3D that were discretized by finite differencing; in the following we limit our discussion to this case.

One traditional way of solving a PDE is to iteratively apply a relaxation algorithm starting from an initial guess of the solution. The Multigrid (MG) algorithm, which is at the heart of the FMG method, builds on relaxation algorithms but improves their convergence rate considerably by temporarily working with a coarser grid than the one on which the solution of the PDE is sought.

The MG algorithm looks essentially like this:

1. *Pre-smooth* the approximate solution of the PDE on a grid at a certain level. Grids are organized into levels, where increasing the level by 1 is equivalent to doubling the resolution of the grid. Smoothing is the term for applying a few sweeps of a traditional relaxation algorithm. This leads to an improved solution of the PDE.

2. *Compute the residual* of the solution. The residual is a measure of the accuracy of an approximate solution and vanishes if the solution is exact.

3. *Restrict* the residual to the grid at the next coarser level. The assumption is that the PDE will be more easily solved on a coarser grid. Restriction is the process of taking a scalar function that is sampled on a grid on a certain level and calculating the values of the function on a grid of the next coarser level.

4. If at coarsest level then *exactly solve* the PDE at this level. This can often be done by direct arithmetic solution. Else, *recursively invoke the MG algorithm* to solve the problem that is defined by the PDE and the residual at the next coarser level. The cycling strategy determines how many recursive calls to the MG algorithm are made at each level.

5. *Interpolate* the solution from the next coarser level. Interpolation is the process of taking a scalar function that is sampled on a grid on a certain level and calculating the values of the function on a grid of the next finer level.

6. *Correct* the solution by taking into account the solution interpolated from the next coarser level. This improves the approximate solution of the PDE.

7. *Post-smooth* the approximate solution of the PDE. This once more improves the approximate solution of the PDE.

The FMG algorithm takes this idea one step further: It starts by solving the PDE on a trivially coarse grid and moves to successively finer grids. The solution of the PDE at each grid level is obtained by employing the MG algorithm at this level.

3 Requirements and Design Decisions

The following lists the most important requirements we formulated for the FMG framework and by which design decisions we met them.

3.1 Performance

Solving a PDE on a grid of approximately the size 100x100x100 (dictated by the biophysics behind the equations) is a compute- and memory-intensive task. An absolute requirement is therefore the employment of an algorithm that is efficient in terms of time and space resources. The FMG algorithm fulfills the first requirement and its implementation in this framework is careful to fulfill the second.

Additionally, Java offers built-in support for multithreading. Since SMP's with several CPU's are quite common in the computational chemistry community, we decided that hardware parallelism should be exploited by the FMG framework through multithreaded implementation of the compute-intensive portions of the FMG algorithm (see section 5 below).

3.2 Extensibility

We demanded that the user be shielded from the intricacies of the FMG algorithm. This is achieved by implementing the FMG algorithm in the form of a framework, which is defined as [2] "a set of cooperating classes that makes up a reusable design for a specific class of software. [...] A developer customizes the framework to a particular application by subclassing and composing instances of framework classes." In particular, the user adapts the FMG framework to a different PDE by implementing the Java interface PDE.

The FMG algorithm involves several choices for sub-algorithms (smoothing, interpolation, restriction; as detailed in section 2) and other details (boundary conditions, cycling strategy) that have to be chosen mostly by trial-and-error for the fastest possible solution of a new PDE. It is essential that the experimentation with these options is made as easy as possible. This is achieved by defining a Java interface for each sub-algorithm and using composition rather than inheritance (according to the Strategy pattern [2]) to combine classes implementing these interfaces to a fully functional FMG algorithm. Javas strong typing makes this approach as secure as it can get.

3.3 Reactivity

Even with the FMG algorithm, solving a PDE in 3D takes some time (see section 7 below). For an applet using the FMG framework it is definitely not acceptable to wait for the return of the method that performs the solution of the PDE for such a long time, because the applet must remain reactive. Therefore we required that the solution of the PDE must itself execute asynchronously in a separate thread. We employ the Observer pattern [5, 2] so that any number of observers can register themselves with the FMG framework and be subsequently notified whenever the solution to the PDE becomes available. The observer pattern is native to Java 1.1 via class java.util.Observable and interface java.util.Observer.

4 An Example

The following short examples should give a feeling for how the FMG framework is used to solve a PDE, namely the Poisson equation $\Delta\Phi(r) = -4\pi\rho(r)$ for a single point-charge in the center of a cubic grid.

All we have to do is implement the interface PDE that characterizes the PDE we want to solve. This is done by implementing the four simple methods evaluateLHS(), evaluate(), sampleRHS(), and getGridSpacing() of this interface (cf. figure 1) which capture the functional form of the PDE and the domain in 3D-space on which the PDE is to be solved. We also define a simple main() method so that this class can run as a standalone Java application.

```
import AT.Ac.univie.imp.loeffler.pde.threeD.fd.*; // the FMG framework
import AT.Ac.univie.imp.loeffler.parallel.*;      // the parallelization package
```

```
public class PoissonPointCharge implements PDE {
    public PoissonPointCharge(double boxLength, double charge) {
        this.boxLength = boxLength; this.charge = charge;
    }
    public double getGridSpacing(int size) {
        return (boxLength/(size - 1.0));
    }
    public double evaluateLHS(ConstBoundaryGrid u, int x, int y, int z) {
        double h = getGridSpacing(u.size());
        return ((1.0/(h*h))*(u.get(x+1,y,z)+u.get(x-1,y,z)+u.get(x,y+1,z)+
                u.get(x,y-1,z)+u.get(x,y,z+1)+u.get(x,y,z-1)-6*u.get(x,y,z)));
    }
    public double evaluate(ConstBoundaryGrid u, ConstNoBoundaryGrid f, int x,
                           int y, int z) {
        double h = getGridSpacing(u.size());
        return ((1.0/6.0)*(u.get(x+1,y,z)+u.get(x-1,y,z)+u.get(x,y+1,z)+
                u.get(x,y-1,z)+u.get(x,y,z+1)+u.get(x,y,z-1)-h*h*f.get(x,y,z)));
    }
    public NoBoundaryGrid sampleRHS(int size) {
        NoBoundaryGrid f = new NoBoundaryGrid(size,0);
        double h        = getGridSpacing(size);
        f.set(size/2,size/2,size/2,-4*Math.PI*charge/Math.pow(h,3));
        return f;
    }
    public static void main(String[] args) {
1       Parallelizer.setDefaultNumberOfThreads(4);
2       PDE          pde = new PoissonPointCharge(100,0.001);
3       Smoother     sm  = new RedBlackGaussSeidel(pde);
4       Restrictor   res = new RestrictorByStencil(RestrictorByStencil.FULL_WEIGHTING);
5       Interpolator in  = new InterpolatorByStencil(InterpolatorByStencil.TRILINEAR);
6       Solver       s   = new ArithmeticSolver(pde);
7       BoundaryGrid g   = new FixedBoundaryGrid(3,0,0);
8       FMG              fmg = new FMG(pde,sm,res,in,s,g,g);
        try {
9           fmg.fmg(1,6,2,2,1,1);
10          ConstBoundaryGrid result = fmg.waitForResult();
            // do something with result
        } catch (Exception e) {return;}
    }
    private double boxLength,charge;
}
```

Examining main() shows that to use the FMG framework one has to decide on how many processors to distribute the calculations (line 1), which PDE to solve (line 2), which smoothing algorithm to use (line 3), which restriction (line 4) and interpolation (line 5) methods to employ, how to solve the PDE at the coarsest level (line 6) and finally which boundary conditions to use (line 7). Using the above ingredients one constructs an FMG object (line 8) and starts the solver (line 9), which runs in its own thread. Since we do not register any observers with the FMG object, we simply wait in line 10 for the result of the problem.

5 Implementation

At the heart of the FMG framework is the concept of a grid, which is in our case a regularly spaced, cubic grid with one floating point value at each grid element. Grids know how to treat their boundary, so that all algorithms using grids are shielded from the detail of which boundary conditions are to be used. For that purpose we defined an abstract class BoundaryGrid that is the superclass of all classes that implement a concrete strategy for handling boundary conditions, like class `FixedBoundaryGrid` for Dirichlet conditions and class `PeriodicBoundaryGrid` for periodic boundary conditions. Every algorithm in the FMG framework works with the highest possible abstraction in the grid hierarchy that still captures the essential properties of the grid. If a method needs to create a new grid it does so by calling the Factory Method [2] `newInstance()` on an existing grid.

Figure 1 shows the relationship of the remaining core types of the FMG framework. We use a notation based on OMT (Object Modeling Technique) [9, 8].

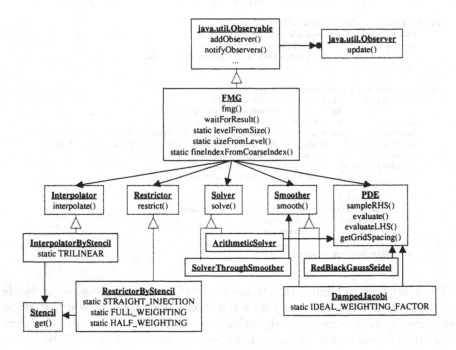

Fig. 1. The type hierarchy of the FMG framework without the grid hierarchy and the types used for parallelization

To implement the Observer pattern, class `FMG` extends class `java.util.Observable` which holds references to any number of interfaces `java.util.Observer`. An observer that would like to be notified when the FMG algorithm has produced a result registers itself with a `FMG` object by calling method `addObserver()`. When the result of the FMG algorithm is available, the observers `update()` method will be called with the result as an argument to this method.

Class `FMG` implements the backbone of the FMG algorithm and delegates the execution of sub-algorithms (see section 2) to interfaces to which it holds references (Strategy pattern [2]). These interfaces are `Interpolator` for the interpolation from a coarse grid to the next finer grid, `Restrictor` for restriction from a fine grid to the next coarser grid, `Solver` for solving the PDE at the coarsest level, `Smoother` for performing one sweep of a smoothing method, and finally `PDE` for capturing the interface to any PDE. We supply at least one concrete implementation (see figure 1) for each of these sub-algorithms, but newly added implementations are guaranteed by Java's strong typing to be compatible with class `FMG`. Those familiar with the FMG algorithm will immediately recognize the popular algorithms that we supply with the FMG framework:

- Interpolation by specification of an interpolation stencil via class `InterpolatorByStencil`.
- Restriction by specification of a restriction stencil via class `RestrictorByStencil`.
- For solving the PDE at the coarsest level we supply class `ArithmeticSolver`, which implements the direct arithmetic solution of the PDE via the methods defined in interface `PDE` and class `SolverThroughSmoother`, which applies any relaxation (smoothing) algorithm until the solution of the PDE is self-consistent.
- Smoothing using either the damped Jacobi algorithm (via class `DampedJacobi`) or the red-black Gauss-Seidel relaxation (via class `RedBlackGaussSeidel`).

Most of the time, solving a new PDE would mean to just implement interface PDE and choose from the already available algorithmic components of the FMG framework — as we have done in the short example above.

All the concrete implementations of the compute-intensive components of the FMG algorithm that we supply with the FMG framework (most of the solid boxes in figure 1) are implemented in a multithreaded fashion, trying to equally partition the total work-load into as many threads as there are processors on the SMP machine. In other words, we are using a very fine-grained parallelism, where parallel execution starts with the invocation of a method and stops at a synchronization point just before the return of the same method [5]. Since Java does not offer a portable way of querying the number of CPU's installed in a machine, the user of the FMG framework has the responsibility of supplying this number. The FMG framework will then distribute the work of compute-intensive algorithms to this number of Java threads and hope that the Java Virtual Machine (JVM, the Java interpreter) and the underlying operating system schedule the threads efficiently and evenly.

6 Multithreading in Java

Java supports multithreading by a direct representation of threads (class `java.lang.Thread`) and by monitors [4] for thread synchronization. Although this paradigm is different from the thread API's (e.g. POSIX threads) usually found on modern operating systems it represents a convenient and portable way of writing multithreaded programs.

Java does not make any assumptions about thread scheduling, though. The standard implementation of the JVM uses so-called 'green threads' which schedules threads on one CPU regardless of the number of CPU's installed in a system. At the time of this writing (September 1997) JavaSoft offers an early-access release of the Java Native Threads Pack (JNTP) for Solaris, the purpose of which is to enable the JVM to use native threads when executing a Java program.

With the help of Sun Austria we were able to perform some preliminary tests of the JNTP on an UltraSPARC SMP. These tests suggest that the JNTP does indeed allow parallel processing on several CPU's of the SMP from a multithreaded Java program, although the early-access release does not yet scale as efficiently as when Solaris threads were directly used. Furthermore, with the current version of the JNTP it is necessary to make a (inherently platform-dependent) call to a native C function to tell the JVM to use all available CPU's. Even with this early-access release of the JNTP the speedup on 2 CPU's was close to the optimum of 2.

Silicon Graphics (SGI) is currently porting the JNTP to their machines but has no preliminary version available yet. However, SGI developed a customized version of the Java-to-C translator TOBA (`http://www.cs.arizona.-edu/sumatra/toba/`) that works with native threads. Unfortunately, TOBA builds on Java 1.0 and not on Java 1.1 and was therefore not directly usable with the FMG framework.

7 Serial Performance

Java has a reputation for being slow. Especially in the area of scientific computing it is the performance of FORTRAN77 programs that is the ultimate benchmark. Therefore we implemented a simple FORTRAN77 version of the FMG algorithm that was restricted to one specific choice for each of the sub-algorithms of the FMG algorithm listed in section 2 (trilinear interpolation, full weighting restriction, red-black Gauss-Seidel relaxation, and arithmetic solving at the coarsest level). We compared the execution time of this FORTRAN77 version to the execution time of the Java version using the same details of the FMG algorithm and employing only one thread. The timings were done on a SGI Challenge with a 150MHz MIPS R4400 CPU, which is a low-end CPU by current standards.

The timings shown in table 1 are based on SGI's FORTRAN77 compiler using "-O" as a compilation flag and SGI's port of JDK1.1.3 using "-O" as a compilation flag as well. This version of the JDK uses a JIT (just-in-time) compiler by default, which translates Java Bytecode into native MIPS machine code as the Java program is executed by the JVM. As can be seen from table 1 the Java FMG framework is approximately 16 times slower than the FORTRAN77 version. The absolute time for solving a PDE with the Java FMG framework is of the order of either 1 or 10 minutes, depending on the size of the grid on which the PDE is discretized.

Table 1. Comparison of Java and FORTRAN77 execution times

grid size	t_{F77}/s	t_{Java}/s	$\frac{t_{Java}}{t_{F77}}$
65x65x65	4.8	78.5	16.4
129x129x129	39.2	612.2	15.6

8 Availability and Acknowledgments

The Java Bytecode of the FMG framework and related packages is freely available from the author (`Gerald.Loeffler@univie.ac.at`). HTML documentation generated by the javadoc utility can be found at the URL `http://www.imp.-univie.ac.at/Loeffler/Javadoc/`.

We would like to thank Sun Austria and especially Alexander Jenewein for their support in testing the JNTP. Furthermore, we would like to thank SGI and especially Hans Böhm and Suresh Srinivas for providing us with their version of TOBA.

This project was funded by the Austrian 'Fonds zur Förderung der Wissenschaftlichen Forschung' under project P10579-MAT.

References

1. W. L. Briggs. *A Multigrid Tutorial*. Society for Industrial and Applied Mathematics (SIAM), Philadelphia, Pennsylvania, 1987.
2. E. Gamma, R. Helm, R. Johnson, and J. Vlissides. *Design Patterns: Elements of Reusable Object-Oriented Software*. Addison-Wesley, Reading, Massachusetts, 1994.
3. W. Hackbusch. *Multi-Grid Methods and Applications*. Springer, New York, 1985.
4. C. A. R. Hoare. Monitors: An operating system structuring concept. *Communications of the ACM*, 17(10):549–557, 1974.
5. D. Lea. *Concurrent Programming in Java: Design Principles and Patterns*. Addison-Wesley, Reading, Massachusetts, 1997.
6. G. Löffler, H. Schreiber, and O. Steinhauser. Calculation of the dielectric properties of a protein and its solvent: Theory and a case study. *J. Mol. Biol.*, 270(3):520–534, 1997.
7. U. Rüde. The multigrid workbench, `http://wwwhoppe.math.uni-augsburg.de/-~scicomp/gsci/software/xwb/xwb.html`, 1997.
8. J. Rumbaugh. The life of an object model: How the object model changes during development. *JOOP*, 7(1):24–32, 1994.
9. J. Rumbaugh, M. Blaha, W. Premerlani, F. Eddy, and W. Lorenson. *Object-Oriented Modeling and Design*. Prentice Hall, Englewood Cliffs, NJ, 1991.

Automatic Binding of Native Scientific Libraries to Java

Sava Mintchev and Vladimir Getov

University of Westminster, London, UK

Abstract. We have created a tool for automatically binding existing native C libraries to Java. With the aid of the Java–to–C Interface generating tool (JCI), the abundance of existing C and Fortran-77 scientific libraries can more easily be made available to Java programmers. We have applied JCI to bind MPI, PBLAS, ScaLAPACK and other libraries to Java. The approach of automatic binding ensures both portability across different platforms and full compatibility with the library specifications. To evaluate the performance of Java code which accesses native libraries, we have run Java versions of parallel benchmarks from the NAS and ParkBench suites. The results obtained on a distributed-memory IBM SP2 machine show the viability of our approach.

1 Introduction

As a programming language, Java has the qualities needed for writing high–performance applications. With the maturing of compilation technology, such applications written in Java will no doubt appear. Since Java is a fairly new language, however, it lacks the extensive scientific libraries of languages like Fortran-77 and C.

The need for access to scientific libraries in Java can be satisfied by: writing new libraries in Java; manually or automatically translating Fortran-77/C library code into Java (*e.g.* with the *f2j* tool [5]); manually or automatically creating a Java wrapper for an existing native Fortran-77/C library. The last approach, in which we are interested, has the obvious advantage of involving the least amount of work, thus dramatically reducing development time. Moreover, it guarantees the best performance results, at least in the short term, because the well-established scientific libraries usually have multiple implementations carefully tuned for maximum performance on different hardware platforms. Last but not least, by applying the software re-use tenet, each native library can be linked to Java with no need for re-coding or translating its implementation.

2 Binding an Existing Native Library to Java

Binding a native library to Java amounts to either dynamically linking the library to the Java virtual machine, or linking the library to the object code produced by a stand-alone Java compiler. At first sight it appears that this should not be a problem, as Java implementations support a *native interface* via which C functions can be called.[1] There are two problems, however. First, native interfaces are convenient when writing new C code to be called from Java, but are inadequate for linking pre-existing native code. This is because Java in general has different data formats than C, and so existing C code cannot be called from Java without modification.

Second, binding a native library to Java raises portability problems. The native interface is not part of the Java language specification [9], and different vendors offer incompatible interfaces. Furthermore, native interfaces are not yet stable and are likely to undergo change with each new major release of a Java implementation[2]. Thus to maintain the portability of the binding one may have to cater to a variety of native interfaces.

2.1 The Java–to–C Interface Generator

To call a C function from Java, we must supply an actual argument in Java for each formal argument of the C function. Unfortunately, the disparity between data layout in the two languages is large enough to rule out a direct mapping in general. Here are some examples:

- Primitive types in C may be of varying sizes, different from the standard Java sizes.
- There is no direct analog to C pointers in Java.
- Multidimensional arrays in C have no direct counterpart in Java.
- C structures can be emulated by Java objects, but the layout of fields of an object may be different from the layout of a C structure.
- C functions passed as arguments have no direct counterpart in Java.

We want to link a large C library (MPI[16], for example) to a Java virtual machine. Because of the disparity between C and Java data types, we are faced with two options:

1. Rewrite the library C functions so that they conform to the particular native interface of our Java VM.
2. Write an additional layer of "stub" C functions which provide an interface between the Java VM (or rather its native interface) and the library.

Software engineering considerations make option (1) a non-starter: it is not our job to tamper with a library supported by others. But option (2) is not very attractive either, considering that a native library like MPI can have more than a hundred accessible functions. The solution is to choose (2), and automate the creation of the additional interface layer.

---

[1] For simplicity we focus on C here, but the main points generalize to Fortran-77, and to other C-linkable languages.

[2] JNI in Sun's JDK 1.1 is regarded as the definitive native interface, but it is not yet supported in all Java implementations by other vendors.

The *Java-to-C interface generator*, or JCI, takes as input a header file containing the C function prototypes of the native library. It outputs several files comprising the additional interface: a file of C stub functions; files of Java class and native method declarations; shell scripts for doing the compilation and linking. The JCI tool generates a C stub function and a Java native method declaration for each exported function of the native library. Every C stub function takes arguments whose types correspond to those of the Java native method, and converts the arguments into the forms expected by the C library function.

As we mentioned in Sec. 1, different Java native interfaces exist, and thus different code may be required for binding a native library to each Java implementation. We have tried to limit the implementation dependence of JCI output to a set of macro definitions describing the particular native interface. Thus it may be possible to re-bind a library to a new Java machine simply by providing the appropriate macros.

2.2 Binding C Libraries (MPI, BLACS, PBLAS)

The largest native library we have bound to Java so far is MPI with over 120 functions [15]. The JCI tool allowed us to bind all those functions to Java without extra effort. Since MPI libraries are standardized, the binding generated by JCI should be applicable without modification to *any* MPI implementation. As the Java binding for MPI has been generated automatically from the C prototypes of MPI functions, it is very close to the C binding. This similarity means that the Java binding is almost completely documented by the MPI-1 standard, with the addition of a table of the JCI mapping of C types into Java types. So far we have bound MPI to two implementations of the Java virtual machine: JDK 1.0.2 [13] for Solaris, and for AIX 4.1 [11]. The MPI implementation we have used is LAM from the Ohio Supercomputer Center [4].

Other libraries written in C for which we have created Java bindings are the *Parallel Basic Linear Algebra Subprograms* (PBLAS) and the *Communication Subprograms* (BLACS). The library function prototypes were taken from the ParkBench 2.1.1 distribution [17]. Table 1 gives the sizes of JCI-generated bindings for several libraries. In addition, there are some 2280 lines of Java class declarations produced by JCI which are common to all libraries.

2.3 Binding Fortran-77 Libraries (BLAS, PB-BLAS, ScaLAPACK)

The JCI tool can be used to generate Java bindings for libraries written in languages other than C, provided that the library can be linked to C programs, and prototypes for the library functions are given in C. We have created Java bindings for several libraries written in Fortran-77: the *Basic Linear Algebra Subprograms* (BLAS Levels 1–3, PB-BLAS) [6], and the *Scalable Linear Algebra Package* (LAPACK, ScaLAPACK) [3]. The C prototypes for the library functions were inferred by *f2c* [8].

The bindings generated by JCI are fairly large in size (see Table 1) because they are meant to be portable and to support different data formats. On a

Table 1. Native libraries bound to Java.

Library	Written in	Size of Java Binding		
		Functions	C lines	Java lines
MPI	C	125	4434	439
BLACS	C	76	5702	489
BLAS	F77	21	2095	169
PBLAS	C	22	2567	127
PB-BLAS	F77	30	4973	241
LAPACK	F77	14	765	65
ScaLAPACK	F77	38	5373	293

particular hardware platform and Java native interface, much of the binding code may be eliminated during the preprocessing phase of its compilation. As our experiments on IBM SP2 machines so far have shown, a negligible amount of time is spent in the binding itself during execution of Java programs.

The use of native numerical code in Java programs is certain to improve performance, as recent experiments with the Java Linpack benchmark [7] and some BLAS Level 1 functions written in C have shown [2, 12]. By binding original native libraries like BLAS, Java programs can gain in performance on all those hardware platforms where the libraries are efficiently implemented.

3 Experimental results

To evaluate the performance of the Java binding to native libraries, we have translated into Java a C + MPI benchmark, the IS kernel from the NAS Parallel Benchmark suite NPB2.2 [1] The program sorts in parallel an array of N integers; $N = 8$ million for IS Class A. The original C and the new Java versions of IS are quite similar, which allows a meaningful comparison of performance results.

We have run the IS benchmark on two platforms: a cluster of Sun SPARC workstations, and the IBM SP2 system at the Cornell Theory Center. Each SP node used has a 120 MHz POWER2 Super Chip processor, 256 MB of memory, 128 KB data cache, and 256 bit memory bus. The results obtained on the SP2 machine are shown in Table 2 and Fig. 1. The Java implementation we have used is IBM's port of JDK 1.0.2D (with the JIT compiler enabled), and the MPI library, a customized version of LAM 6.1[3] We opted for LAM rather than the proprietary IBM MPI library because the version of the latter available to us does not support the re-entrant C library required for Java [10]. The results for the C version of IS under both LAM and IBM MPI are also given for comparison.

It is important to identify the sources of the slowdown of the Java version of IS with respect to the C version. So we instrumented the JavaMPI binding, and

---

[3] Earlier results obtained with the original LAM 6.1 as reported in [15] show poor scalability.

Table 2. Statistics for the C and Java IS benchmarks on an IBM SP2.

Class	Language	MPI implement	Execution Time (sec)					Mop/s Total				
			1	2	4	8	16	1	2	4	8	16
A	Java	LAM	—	48.04	24.72	12.78	6.94	—	1.75	3.39	6.56	12.08
	C	LAM	42.16	24.52	12.66	6.13	3.28	1.99	3.42	6.63	13.69	25.54
	C	IBM MPI	40.94	21.62	10.27	4.92	2.76	2.05	3.88	8.16	14.21	30.35

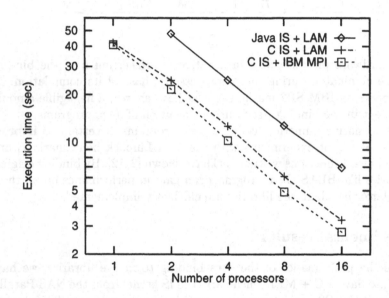

Fig. 1. Execution time for IS Class A on an IBM SP2

gathered additional measurements. It turns out that the cumulative time spent in the C functions of the JavaMPI binding is approximately 20 milliseconds in all cases, and thus the binding has a negligible part of the total execution time for the Java version of IS. Clearly the JavaMPI binding does not introduce a noticeable overhead in the results from Table 2.

Further experiments were carried out with a Java translation of the MAT-MUL benchmark from the ParkBench suite [18, 17]. The original benchmark is in Fortran-77 and performs dense matrix multiplication in parallel. It accesses the BLAS, BLACS and LAPACK libraries included in the ParkBench 2.1.1 distribution. MPI is used indirectly through the BLACS native library. We have run MATMUL on a SPARC workstation cluster, and on the IBM SP2 machine at Southampton University (66MHz Power2 "thin1" nodes with 128Mbyte RAM, 64bit memory bus, and 64Kbyte data cache). The results are shown in Table 3 and Fig. 2.

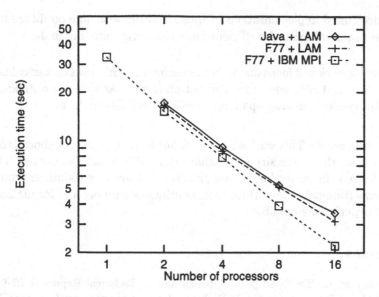

Fig. 2. Execution time for MATMUL (N = 1000) on an IBM SP2.

Table 3. Statistics for Fortran and Java MATMUL benchmarks on an IBM SP2.

Problem size (N)	Lang	MPI implement	Execution Time (sec)					Mflop/s Total				
			1	2	4	8	16	1	2	4	8	16
	Java	LAM	—	17.09	9.12	5.26	3.53	—	117.0	219.4	380.2	566.9
1000	F77	LAM		16.45	8.61	5.12	3.13		121.6	232.3	390.4	638.3
	F77	IBM MPI	33.25	15.16	7.89	3.91	2.20	60.16	132.0	253.6	511.2	910.0

Figure 2 shows that Java MATMUL execution times are only 5–10% longer than Fortran-77 times. These results may seem surprisingly good, given that Java IS is two times slower than C IS (Fig. 1). The explanation is that in MATMUL most of the performance-sensitive calculations are carried out by the native library routines (which are the same for both Java and Fortran-77 versions of the benchmark). In contrast, IS uses a native library (MPI) only for communication, and all calculations are done by the benchmark program.

The performance results from Fig. 2 are a persuasive argument for our approach of linking native scientific libraries to Java.

4 Conclusion

In this paper we have summarised our work on high-performance computation in Java. We have written a tool for automating the creation of portable interfaces to native libraries (whether for scientific computation or message passing). We have applied the JCI tool to create Java bindings for MPI, BLAS, LAPACK, and so forth, which are fully compatible with the library specifications. With performance-tuned implementations of those libraries available on different machines, the potential exists for efficient numerical programming in Java.

Our future work will focus on further experiments with Java numerical benchmarks on the IBM SP2 and other parallel platforms, as well as on making the PMPI [14] high-level message-passing interface available in Java.

Acknowledgments This work was carried out as part of our collaboration with colleagues from the University of Southampton (U.K.) and the Cornell Theory Center (U.S.A.). In particular, we are grateful to Tony Hey (Southampton) and Susan Flynn Hummel (Cornell) for their continuous support and for making the IBM SP2 experiments possible.

References

1. D. Bailey et al. The NAS parallel benchmarks. Technical Report RNR-94-007, NASA Ames Research Center, 1994. http://science.nas.nasa.gov/Software/NPB.
2. A.J.C. Bik and D.B. Gannon. A note on native Level 1 BLAS in Java. In *[12]*, 1997.
3. L.S. Blackford, J. Choi, A. Cleary, E. D'Azevedo, J. Demmel, I. Dhillon, J. Dongarra, S. Hammarling, G. Henry, A. Petitet, K. Stanley, D. Walker, and R.C. Whaley. ScaLAPACK: A linear algebra library for message-passing computers. In *SIAM Conference on Parallel Processing*, 1997.
4. G. Burns, R. Daoud, and J. Vaigl. LAM: An open cluster environment for MPI. In *Supercomputing Symposium '94*, Toronto, Canada, June 1994. http://www.osc.edu/lam.html.
5. H. Casanova, J.J. Dongarra, and D.M. Doolin. Java access to numerical libraries. In *[12]*, 1997.
6. J. Choi, J. Dongarra, and D. Walker. PB-BLAS: A set of parallel block basic linear algebra subroutines. In *Proceedings of the Scalable High Performance Computing Conference, Knoxville, TN*, pages 534–541. IEEE Computer Society Press, 1994.
7. J. Dongarra and R. Wade. Linpack benchmark – Java version. http://www.netlib.org/benchmark/linpackjava.
8. S. I. Feldman and P. J. Weinberger. *A Portable Fortran 77 Compiler. UNIX Time Sharing System Programmer's Manual, Tenth Edition*. AT&T Bell Laboratories, 1990.
9. J. Gosling, W. Joy, and G. Steele. *The Java Language Specification, Version 1.0*. Addison-Wesley, Reading, Mass., 1996.
10. IBM. *PE for AIX: MPI Programming and Subroutine Reference*. http://www.rs6000.ibm.com/resource/aix_resource/sp_books/pe/.
11. IBM UK Hursley Lab. *Centre for Java Technology Development*. http://ncc.hursley.ibm.com/javainfo/hurindex.html.
12. *ACM Workshop on Java for Science and Engineering Computation*, Las Vegas, Nevada, June 21 1997. To appear in *Concurrency: Practice and Experience*. http://www.cs.rochester.edu/u/wei/javaworkshop.html.
13. JavaSoft. Home page. http://www.javasoft.com/.

14. S. Mintchev and V. Getov. PMPI: High-level message passing in Fortran77 and C. In Bob Hertzberger and P. Sloot, editors, *High-Performance Computing and Networking (HPCN'97)*, pages 603–614, Vienna, Austria, 1997. Springer LNCS 1225.

15. S. Mintchev and V. Getov. Towards portable message passing in Java: Binding MPI. In *Proceedings of EuroPVM-MPI*, Kraków, Poland, November, 1997. To appear in Springer LNCS.

16. MPI Forum. MPI: A message-passing interface standard. *International Journal of Supercomputer Applications*, 8(3/4), 1994.

17. PARKBENCH Committe. Parallel kernels and benchmarks home page. http://www.netlib.org/parkbench.

18. PARKBENCH Committe (assembled by R. Hockney and M. Berry). PARK-BENCH report - 1: Public international benchmarks for parallel computers. *Scientific Programming*, 3(2):101–146, 1994.

JAPE : The Java Parallel Environment

Kenji Imasaki

School of Computer Science, Carleton University,
Ottawa, Ontario K1S5B6, Canada

Abstract. This paper presents the design and implementation of a JAva Parallel Environment (JAPE) which is based on JPVM. Its goals are to provide Java with full PVM message-passing functions and to improve the performance of JPVM. Particularly, the improvements of JPVM by parallel task spawning and message-packed broadcast are described. These improvements are tested on benchmark programs. The results show they have favorable effects in the case where one processor has many tasks. The design and implementation of a graphical user interface to JAPE are also described.

1 Introduction

As the computer and network technology advances, Networks Of Workstations (NOWs), in which independent computers with their own CPUs, disks and memories are connected via a high-speed network, can be used to solve large scientific problems. For such NOWs, PVM [2] (Parallel Virtual Machine) and MPI [3](Message Passing Interface) are the most popular message-passing programming languages for distributed systems. Meanwhile, Java, SunSoft's architecture independent object-oriented programming language for distributed systems, is becoming popular. Java provides good user-interfaces and network facilities. However, Java does not provide basic message-passing interfaces needed to program message-passing parallel computing; send, receive, synchronization and parallel reduction among processors are not implemented in Java.

Therefore, there has been growing interest in integrating PVM or MPI and Java. HPJava [5], JPVM [1] and JavaPVM [6] are being designed and implemented. Among these projects, JPVM attempts to accomplish the same goals as our project.

Another aspect of parallel programming is a parallel debugger/profilier. When programmers want to develop parallel programs, profilers becomes important for improving program performance and parallel debuggers help programmers. However, Java does not provide any profiler/debugger with a user-friendly interface for parallel computing.

Hence, the goals of the project, JAva Parallel Environment (JAPE) [4], which is based on JPVM, are :

- Provide Java with full message-passing PVM functions;
- Improve performance of JPVM;
- Provide Java with a user-friendly parallel profiler/debugger(JAPEView).

2 Related Work

Currently, several projects are being carried out to accomplish parallel computing in Java.

HPJava [5] is being designed by cooperation of several universities such as
Indiana University and Harvard University. The on-going discussion of HPJava
ranges from possible applications of parallelism of Java to Java language extensions for parallel processing.

Other interesting projects are JavaPVM [6] and JPVM [1], which attempt to
integrate Java and PVM.

JavaPVM is an interface written with the Java native methods capability
which allows Java applications to use PVM. HPJava also may adopt the same
approach. JAPE didn't take this approach because converting data from Java
to PVM will be considered overhead as the execution of Java programs becomes
as fast as that of PVM or MPI in the future.

On the other hand, JPVM is an implementation of PVM written in Java.
The message sending function (pvm_send), for example, is implemented by using ONLY Java network facilities. However, only part of PVM functions have
been implemented in JPVM. For instance, one of the important functions in
parallel computing – the reduction function (pvm_reduction) – has not been
implemented yet. In addition, its performance is not so good as PVM.

3 JAPE's Improvements on JPVM

JPVM has some potential improvement aspects. In particular, JAPE tries to
improve the task creation and broadcast processes in JPVM.

3.1 Task Creation(pvm_spawn())

The improvement of task creation lies in the parallelization of task creation
processes.

In JPVM, the spawn daemon asks the other daemons to invoke a certain
number of user-specified tasks in a round-robin way. It should be noted that
the original JPVM does this sequentially; first, it asks one daemon to invoke a
task and then asks another after getting the task creation completed message
from the first daemon. The task creation completed message contains the port
number of the task.

In order to speed up pvm_spawn, parallelization of task creation processes
is the best way. To parallelize it, the spawn daemon asks another daemon to
invoke a task without waiting for the task creation completed message. Each task
is created independently and sends the task creation completed message back
to the spawn daemon. After sending all task invocation messages, the spawn
daemon waits for all task creation completed messages from the other daemons.

3.2 Broadcast(pvm_mcast())

In PVM, pvm_mcast is to used to send a message to all the tasks specified in its argument.

JPVM implements pvm_mcast by establishing socket connections from the source task to all the destination tasks (they are shown in broken lines in Fig. 1). However, because the messages to be sent are the same, a task can send a message just once to each computer instead of sending messages to all destination tasks (they are shown in solid lines in Fig. 1). This improvement can reduce the number of remote messages dramatically when there are many tasks on a single computer. In other words, the messages to be sent to tasks on the same computer are *packed* into one message. In detail, it is implemented as follows :

1. task invoking pvm_mcast sends a message and a task list to the local daemon;
2. the daemon receiving the message separates the task list according to hostnames and sends the message with a task list to each daemon;
3. the daemon receiving the message sends the message to the tasks in the task list.

This implementation actually increases the execution steps and the number of all messages because the task and daemon communicates with jpvm_send even if they are on the same computer; however, this local communication is much faster than remote communication. If the number of tasks on a computer is large, these efforts are considered to be effective.

4 JAPEView

JAPEView provides a graphical interface to the JAPE console commands and the above information, along with several animated views to monitor the execution of JAPE programs. These views provide information about the interactions among tasks in a parallel JAPE program, to assist in debugging and performance tuning.

An example of an execution of JAPEView is shown in Fig. 2. The screen of JAPEView consists of the console commands, space-time view, network view, and tasks output views. The implementation of JAPEView can be divided into slave side and master side as shown in Fig. 3.

4.1 JAPEView Slave Side

The slave side of JAPEView consists of the class StatusSend and MsgSend which reside on each task as shown in Fig. 3.

Whenever the status of a task changes, StatusSend records the change. At certain intervals, it sends them to RecvTask on the master side using pvm_send in a packed form. Whenever a task receives a message from other tasks, MsgSend records the message information such as its source task(st in Fig. 4), destination task(rt), sending time(sm), and receiving time(rm). At certain intervals, it sends

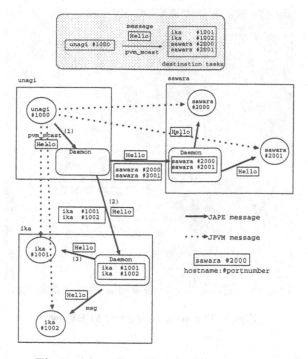

Fig. 1. An implementation of pvm_mcast

them to RecvTask using pvm_send in a packed form. Note that the clock time used here is based on the Java system function currentTimeMillis() which is ideally accurate on all computers.

4.2 JAPEView Master Side

The master side of JAPEView consists of the classes MainScreen and RecvTask which reside on the master host as shown in Fig. 3.

RecvTask receives messages about statuses and message information from StatusSend and MsgSend, respectively, as shown in Fig. 3. The received data is stored in staus_table and msg_table, respectively. Besides, the contents of both tables are stored in the user-specified file for users' convenience. RecvTask also receives outputs from all tasks.

MainScreen, which is implemented by an independent thread, continuously updates the task image and message graphs based on both tables as shown in Fig. 4. The task output view shows the outputs from all tasks.

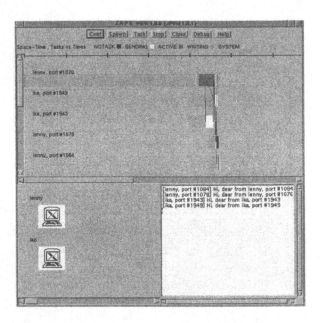

Fig. 2. The main screen of JAPEView

5 Evaluation of Improvements

The improvements described in Section 3 were tested on NOWs. Table 1 shows the configuration of the computers. Note that the elapsed time shown here is the average of five execution times.

Table 1. Configuration used in this evaluation

OS	LINUX 2.0.28
CPU	P150
Memory	32MByte
Java Compiler	JDK 1.02pl2 (without -O option)
Network	Ethernet

5.1 Parallel Spawn

In this benchmark program to test the algorithm in Section 3.1, first, the master task invokes a certain number of tasks by **pvm_spawn**. Then, the invoked tasks

142

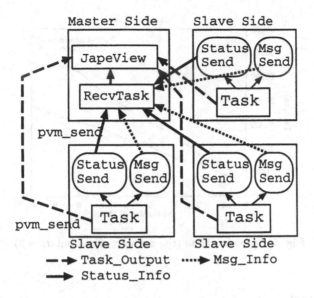

Fig. 3. Two Components of JAPEView

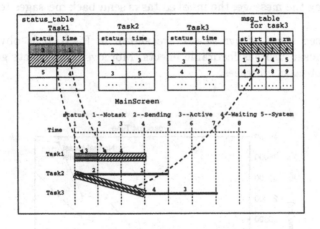

Fig. 4. JAPEView Master Side

simply send messages back to the master task. Finally, the master task receives these messages and terminates.

The elapsed times of this program are shown in Fig. 5. It is obvious that parallel spawning has favorable effects. Moreover, the effects get better as the number of tasks increases.

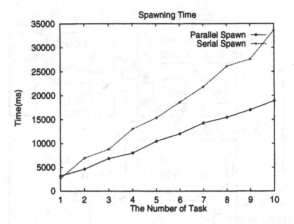

Fig. 5. Spawning time (the number of computers is 2)

5.2 Broadcast

In this benchmark program to test the algorithm in Section 3.2, first, the master task invokes tasks. Then, the master task broadcasts a message by pvm_mcast. After receiving the message, the invoked tasks send back messages to the master task.

The elapsed times of pvm_mcast are shown in Fig.6. It is obvious that a packed broadcast also has favorable effects. Moreover, the effects get better as the number of tasks increases.

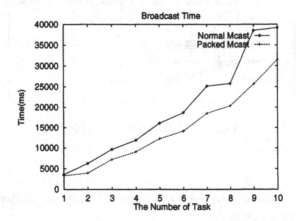

Fig. 6. pvm_mcast Result (the number of computers is 2)

6 Conclusion

This paper presents the design and implementation of a JAva Parallel Environment (JAPE) which is based on JPVM. The difference between JAPE and JPVM is parallel task spawning and a message-packed broadcast. These improvements are tested by benchmark programs. The benchmark results show that these improvements are effective when one processor has many tasks. In addition, the effects get better as the number of tasks increases. However, further optimization of pvm_spawn (e.g. use of broadcast) is needed to use JAPE for real parallel applications. This paper also presents the design and implementation of a graphical interface to JAPE, JAPEView. It provides a graphical interface to the JAPE console commands and various information, along with several animated views to monitor the execution of JAPE programs. JAPEView also facilitates a user friendly way to start a parallel program.

References

1. A. J. Ferrari. JPVM : The Java Parallel Virtual Machine. http://www.cs.virginia.edu/ ajf2j/jpvm.html.
2. A. Geist, A. Beguelin, J. Dongarra, W. Jiang, R. Manchek, and V. Sunderam. *PVM: Parallel Virtual Machine A Users' Guide and Tutorial for Networked Parallel Computing.* MIT Press, 1994.
3. W. Gropp, E. Lusk, and A. Skjellum. *Using MPI: Portable Parallel Programming with the Message-Passing Interface.* MIT Press, 1994.
4. K. Imasaki and Z. Gu. JAPE : Java Parallel Envornment. http://www.cs.mcgill.ca/ imasaki/report/report.html.
5. The Parallel Compiler Runtime Consortium (PCRC). HPCC and Java. http://www.npac.syr.edu/users/gcf/hpjava3.html.
6. D. A. Thurman. JavaPVM : The Java to PVM interface. http://www.isye.gatech.edu/chmsr/JavaPVM/.

An Architecture in Java for Mobile Computation

Ju-Pin Ang and Yong-Tai Tan

DSO National Laboratories,
20 Science Park Drive,
Singapore 118230

Abstract. The components required to effectively use a mobile computing infrastructure include the *mobile computing languages* (MCLs), the *programming paradigms* for mobile computation and the *architectures* within which to implement user tools and applications. There are few architectures that exist presently in which users can implement their own tools and applications using a programming paradigm of their choice.

In this paper, an architecture built using Java in which users can create their own tools and applications is presented. The key feature of this architecture is that it is general enough to allow users to implement their own tools and applications without being restricted as to which programming paradigm is used.

Two tools were developed to illustrate the general-purpose nature of the architecture. The first is a simple network information tool which supplies information about a node in the network when queried. The second tool performs a distributed branch-and-bound search of a tree structure. This second tool gives a glimpse of the possibility of implementing distributed high performance computing solutions on this architecture.

1 Introduction

There has been increasing interest and research on mobile computation due to the availability of a widespread communication infrastructure in the the form of the Internet [?]. One view of the mobile computation infrastructure is that of a distribution of computational environments (CEs) which support execution units (EUs). The components required to effectively use such a mobile computational infrastructure include the *mobile computing languages* (MCLs), the *programming paradigms* for mobile computation and the *architectures* within which to implement user tools and applications. Mobile computing languages form the basis of any implementation and are being developed by groups both in academia and industry. One of the most well-known mobile computing languages is Java from Sun Microsystems [?]. There are also groups which are formalising the programming paradigms, for example, Carzaniga et. al. [?]. However, there are few architectures which exist presently in which users can implement their own tools and applications using a programming paradigm of their choice.

In this paper, an architecture in which users can create their own tools and applications is presented. It is built using Java as the mobile computing language and hence supports weak mobility. The key feature of this architecture is that it is general enough to allow users to implement their own tools and applications without being restricted as to which programming paradigm is used. In this architecture, users can conceivably use the Client-Server, Remote Evaluation, Code-on-Demand as well as Mobile Agent paradigms. This is in contrast to existing systems such as MOLE [?] for example, which is tailored for development using the Mobile Agent paradigm.

Two tools were developed to illustrate the general-purpose nature of the architecture. The first is a simple network information tool which supplies information about a node in the network when queried. This tool is implemented using a hybrid Client-Server, Remote Evaluation paradigm. The second tool performs a distributed branch-and-bound search of a tree structure. This is an example of using the Remote Evaluation paradigm. Algorithms involving branch-and-bound are widely used in AI and this second tool gives a glimpse of the possibility of implementing distributed high performance computing solutions on this architecture.

This paper is structured as follows. Section 2 describes the proposed architecture and gives details about its implementation. In Section ??, the two tools mentioned above are discussed in more detail. In Section ??, the major points of this paper are summarised and conclusions presented.

2 The mobile computing architecture

The architecture consists of three layers and is shown in Figure 1. The base layer, called the *Distributed Objects Layer*, implements distributed objects and the mechanisms for their communication. This layer consists of the Java language with extensions for Remote Method Invocation (RMI) in the form made available by the Java Development Kit 1.1. There are several conveniences gained by using Java. Firstly, serialisation of the distributed objects is handled by making them implement the java.io.Serializable interface. Secondly, communication between objects can be implemented using the RMI mechanisms and naming of the objects can be handled by the java.rmi.Naming class. Finally, the Java security model can be relied upon as a first step in handling security issues.

The middle layer, called the *Agent Layer*, consists of distributed objects which implement a minimal set of methods which allows them to be enrolled in an RMI registry. These distributed objects, or *Agents*, correspond to the execution units of the mobile computing infrastructure. Although they have been called Agents, these distributed objects can be used to implement a tool or application using a paradigm chosen by the user, not necessarily the Mobile Agent paradigm. A special kind of Agent, called the *AgentManager*, resides on each of the network nodes and is responsible for managing the Agents resident on its network node. The AgentManager also communicates with a JDBC-connectable database to maintain details of its Agents persistently. By using JDBC, the choice of database implementation is not restricted provided each database follows the same schema. This layer is described fully in Section 2.1.

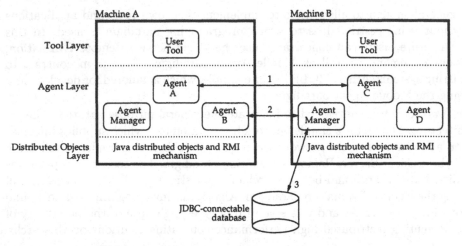

Notes:
1. Agents can communicate with each other though remotely-invokable methods.
2. The Agent Manager performs its management operations by invoking certain RMI methods available on every Agent.
3. The Agent Manager saves management information about the agents on a JDBC connectable database.

Fig. 1. Architecture for mobile computing.

The top layer, called the *Tool Layer*, consists of *tools* or applications built using the Agents of the Agent Layer. A tool forms the interface between the user and the Agents residing on the network nodes. It initiates the distribution of the Agents if necessary and presents the results of the mobile computation to the user.

2.1 The Agent Layer

The class and interface hierarchies of the distributed objects making up the Agent Layer are shown in Figure 2. An Agent consist of two parts – the interface, which declares the remotely invokable methods; and the implementation, which implements the methods declared in the interface as well as undeclared methods. Thus, there is a one-to-one correspondence between an interface and its implementation class. All Agent classes extend or inherit from the Agent-Object class. The AgentObject class implements a start() method which enrolls the Agent with the RMI registry of the local node.

The AgentManager class is such an extension of the AgentObject class. Additionally, it implements a main() method to allow it to be started from a script or from the command line. The main() method starts a local RMI registry and if requested, can load all known Agent objects pertinent to the local node which have their details stored in the JDBC-connectable database. The notion that the AgentManager extends an AgentObject means that newer versions of Agent-

Class Hierarchy

```
... java.rmi.server.UnicastRemoteObject
    └─ NetTool.AgentObjectImpl (implements AgentObject, Serializable)
        ├─ NetTool.AgentManagerImpl (implements AgentManager, Serializable)
        ├─ NetTool.PathInfoImpl (implements PathInfo, Serializable)
        ├─ NetTool.InfoServerImpl (implements InfoServer, Serializable)
        └─ ... other Agent classes
```

Interface Hierarchy

```
... java.rmi.Remote
    └─ NetTool.AgentObject
        ├─ NetTool.AgentManager
        ├─ NetTool.PathInfo
        ├─ NetTool.InfoServer
        └─ ... other Agent interfaces
```

Fig. 2. Class and Interface hierarchy for Agent Layer.

Manager may be loaded by an existing AgentManager object to replace itself, provided that the class loader does not cache the existing AgentManager class.

The AgentManager accepts several input parameters when started, namely the location of the database, JDBC driver, the registry port and path location of Agent classes. These settings are stored in a PathInfo object so that they are persistent when an AgentManager is replaced as described above. The AgentManager class also overides the start() method of AgentObject because, in addition to all of the operations implemented in the start() method of AgentObject, it has to load the configuration settings from the PathInfo object .

Two overloaded methods, called loadAgent(agentName) and loadAgent(URL, agentName), are provided by AgentManager to load Agent classes either from the local file system or from another network node respectively. Both methods retrieve the specified Agent class, instantiate an Agent object and use the start() method of the Agent object to enroll it in the RMI registry. Details about the original location of the Agent class are stored in the JDBC-connectable database so that when AgentManager is started in future, previously loaded Agents can be located and can be loaded if requested.

For developers other than the author of an Agent to use the methods provided by that Agent, they must be able to view an Agent interface. In addition, the compiled Agent interface must the available in a developer's classpath for successful compilation of code that utilises the Agent's methods. For this purpose, AgentManager has a complementary set of methods, called URL makeInstallableAgentIntf(agentName) and installAgentIntf(URL, agentName). The method makeInstallableAgentIntf(...) packages an Agent's interface source and compiled

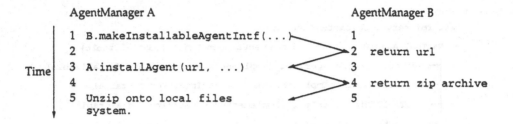

Fig. 3. Installing an Agent interface.

class files in a zip archive. The returned URL gives the location from which this zip archive can be retrieved. The installAgentIntf(...) method retrieves the zip archive from the given URL and unzips it onto the local file system. Figure **3.** shows this process. It is arguable whether such an operation should be permitted given its security implications. However, in the current closed operating environment comprising of an isolated LAN where the developers are non-malicious, this operation is allowed without further checks. In an open environment, some form of trust relationship will have to be established and data integrity will have to be ensured before this operation can be allowed.

To reduce network traffic, one solution is for the Agent classes to be cached on a network node. This avoids having to load an Agent from across the network whenever an AgentManager is started. A complementary set of methods similar to URL makeInstallableAgentIntf(agentName) and installAgent(URL, agentName) are provided. URL makeInstallableAgentImplStubSkel(agentName) and installAgentImplStubSkel(URL, agentName) create a zip archive of the Agent class and unzip an archive from the supplied URL respectively. As with URL makeInstallableAgentIntf(agentName) and installAgentIntf(URL, agentName), these operations are currently allowed without further checks in the closed environment.

The methods provided by AgentManager are summarised in Table **1.** Using these methods, users can create their own tools by writing Agents and distributing them across the network.

3   Examples of tools

Two tools were developed using this architecture. The first one is a simple network information tool. The second implements a distributed branch-and-bound search of a tree structure. Each of these is discussed in detail in the following sections.

3.1   Simple Network Information Tool

This tool makes use of an Agent, called an InfoServer, which provides remotely invokable methods for obtaining information about the network node on which

Table 1. Remotely invokable methods of `AgentManager`.

Method	Description
loadAgent(url, agentName)	Load an Agent class from a specified URL, instantiate to create an Agent object and register it in the RMI registry.
loadAgent(agentName)	Load an Agent class from the local file system, instantiate to create an Agent object and register it in the RMI registry.
URL makeInstallableAgentIntf(agentName)	Package an Agent interface as a zip archive and return the URL from which it can be retrieved
installAgentIntf(url, agentName)	Retrieve an Agent interface zip archive from the specified URL and unzip it onto the local file system.
URL makeInstallableAgent(agentName)	Package an Agent class as a zip archive and return the URL from which it can be retrieved
installAgentImplStubSkel(url, agentName)	Retrieve an Agent class zip archive from the specified URL and unzip it onto the local file system.

it resides. The methods return information about free memory, total memory, OS name, OS version and OS architecture. The tool is a class called InfoTool with a main() method which first locates the AgentManager object on the node of interest, then loads the InfoServer agent onto that node by calling the loadAgent(...) method. An instance of the InfoServer is then available, and its methods can be used to obtain information about the node.

This tool is an example of a tool developed using a hybrid Client-Server and Remote Evaluation paradigm. The client, in this case, is the InfoTool which is serviced by the InfoServer. Remote Evaluation is used because the InfoServer is sent to the remote node before it begins its service.

3.2 Distributed Branch-and-Bound Tool

This tool implements a distributed branch-and-bound search of a tree structure. Each node of the tree consists of a unique tuple of values and has an associated cost dependent on the values in the tuple. The search aims to identify the node with the lowest cost. In addition, certain subtrees can be pruned if the cost at the root of the subtree already exceeds the lowest cost discovered so far. This kind of problem occurs frequently in AI, where the tuples may contain integers, characters, strings or data structures representing more complex information such as game moves.

Such trees have characteristics which provide convenient abstractions for distributing the search. Firstly, each subtree represents a complete computation since there is enough information at the root of a subtree to perform the expansion and search of that subtree. Secondly, each tuple of the tree constitutes one unit of work since there is enough information in a node to evaluate the cost. A distributed search then involves the distribution of the tuples and associated subtrees.

The tool makes use of an Agent, called BnbAgent, which provides four remotely invokable methods: giveWork(), doWork(), seedWork(...) and addHosts(...). An instance of BnbAgent resides on each machine taking part in the search. Every BnbAgent has a queue for each level of the tree to store the tuples or units of work. These work queues are of a fixed length, so memory use is controlled and swapping behaviour can be avoided. To accommodate searches where there are a large number of tuples in any level of a tree, a work queue is replenished by generating more tuples from the last node in the queue if it becomes empty during the search.

Each BnbAgent repeatedly removes one unit of work from the deepest non-empty work queue and computes the cost of that tuple. If no pruning can occur, its children are generated in the work queue of the next depth. This is performed in doWork(). When a BnbAgent cannot find any tuples in its work queues, it steals one unit of work from another BnbAgent's shallowest non-empty queue. This is done by invoking the giveWork() method on the BnbAgent from which work is to be stolen. This means that entire subtrees are stolen, minimising the communication/computation ratio. To start the search, the root of the tree is seeded on any BnbAgent. This is performed by seedWork(...). Work stealing then ensures that the idle BnbAgents steal work from the seeded BnbAgent. This results in dynamic load balancing since whenever a BnbAgent becomes idle, it will attempt to steal work from another BnbAgent.

Each BnbAgent keeps its own copy of the lowest cost discovered so far. A BnbAgent may optionally broadcast its lowest cost if it is significantly lower than the previous cost. Broadcasting the cost aids pruning, while doing so only on discovery of a significantly lower cost reduces unnecessary communication.

The work queues and the lowest cost bounds store the entire state of the computation. Thus, checkpointing is possible by storing the work queues and lowest cost bounds persistently.

The use of work stealing means that machines may be added dynamically to the search. The added machine will have empty work queues and will attempt to steal work from other BnbAgents. The other BnbAgents have to be informed of the new BnbAgent to enable them to steal work from it. This done using the addHosts(...) method. Conversely, a node may be removed from the search, provided that all its work queues are empty. However, as yet no methods have been provided to implement this in the current prototype.

The entire search is complete when all the work queues belonging to every BnbAgent is empty. At this time, the lowest cost and its corresponding tuple can be presented by the tool to the user.

The tool correctly traversed a tree structure on a heterogeneous pool of three nodes consisting of one Win95 Pentium 100 machine, one WinNT PentiumPro 200 machine and one Solaris 2.5 SPARC 20 machine, with the initial work seeded on any one of the machines.

4 Conclusion

A three-layered architecture for mobile computation has been presented. The architecture comprises of distributed objects at the base layer, Agents at the middle layer and user developed tools at the top layer. The architecture is general-purpose in the sense that the Agents can perform any task within the constraints of the Java security model. As a result, users are not restricted to any particular programming paradigm when writing their own Agents. Users create their own tools by distributing the Agents across the network.

In addition, this architecture can potentially be used to develop distributed high performance computing solutions as seen from Section ??. A distributed, dynamic, work-stealing, heterogeneous, branch-and-bound tree search tool was successfully implemented on this architecture.

Currently, this architecture operates in a closed environment where the developers are non-malicious. As such, security relies on the Java security model. In an open environment, further security issues such as the establishment of trust relationships and guarantee of data integrity will need to be addressed.

Special thanks to Lim Chu Cheow and Christopher Ting with whom the authors had valuable discussions.

References

1. Carzaniga, A., Picco, G. P., Vigna, G.: Designing Distributed Applications with Mobile Code Paradigms. In Proceedings of the 19th International Conference on Software Engineering, Boston, MA, May 1997.
2. Ghezzi, C., Vigna, G.: Mobile Code Paradigms and Technologies: A Case Study. In Proceedings of the First International Workshop on Mobile Agents, Berlin, Germany, April 1997.
3. Straßer, M., Baumann, J., Hohl, F.: Mole - A Java Based Mobile Agent System. In Proceedings of the 2nd ECOOP Workshop on Mobile Object Systems, Linz, Austria, July 1996.
4. Sun Microsystems, Inc. JDK 1.1 Documentation. Mountain View, CA, 1996.

The Extensible Java Preprocessor Kit and a Tiny Data-Parallel Java

Yuuji Ichisugi and Yves Roudier, STA Fellow

Electrotechnical Laboratory, Japan

Abstract. We describe the extensible Java preprocessor EPP and a data-parallel extension of Java implemented with EPP. EPP can be extended by incorporating EPP plugins. These plugins are programmed with the Ld-2 language that we also describe. Tiny Data-Parallel Java is an example of EPP plugin. High portability is guaranteed because the translated code and the run-time systems are pure Java code. Applications can be executed in parallel if the VM interpreter supports parallel execution of Java threads. We provide a preliminary performance evaluation of this system.

1 Introduction

The Java language [6] has recently become very popular among programmers. Java has socket and thread libraries which can be used on various platforms. Many JAVA VM (virtual machine) interpreters will soon support shared-memory multiprocessing and will then be useful platforms for parallel programming.

However, Java lacks facilities for language extension, which have been adopted by other object-oriented languages. For example, C++ has a macro preprocessor, operator overloading and template facilities. Smalltalk and CLOS[15] have closures and metaclass facilities. These features to extend language constructs and operators supplement the inheritance mechanism which extends data types. Without such extension possibilities, it is difficult to make parallel libraries which can be used easily by application writers.

To enhance Java in this regard, we developed an extensible Java preprocessor kit, EPP [8]. EPP can be used to introduce new language features, possibly associated with new syntax.

In this paper, we describe Tiny Data-Parallel Java, an extension that is implemented as an example application of EPP. It is based on the same language model as Data-Parallel C [7]. The source code of a Tiny Data-Parallel Java application is translated to standard Java code by a translator implemented using EPP. Because the run-time system is written in standard Java, high portability is guaranteed.

Tiny Data-Parallel Java does not have enough language features to support high-performance parallel programs; however, we think it shows convincingly the effectiveness of the combination of the high extensibility of EPP and of the high portability of Java, and their value for parallel programming. Many data-parallel extensions are currently being proposed for Java ([14, 13], for example), but none of these frameworks has been introduced by means of a general extension mechanism.

2 An Extensible Preprocessor Kit: EPP

In this section we review existing tools, introduce the Extensible Preprocessor Kit, and explain its programming with the Ld-2 language..

2.1 Existing tools

To implement new languages or extend existing ones, preprocessors or translators are often used rather than native compilers. Many language extensions and source level optimization tools for C/C++ are implemented as translators to C/C++. Recently, several extensions for the Java language have also been proposed which are implemented as preprocessors. Because of the simplicity of Java, it is also relatively easy to provide extension support libraries.

The merits of this style of implementation are its simplicity and high portability. Instruction-level optimization can be delegated to the compiler of the target language.

Although there are potentially many useful language extension systems, a user must select just one for a project, because it is generally impossible to merge several language extensions or eliminate harmful features from the extended system. Systems with a meta-object protocol (MOP) such as CLOS [11], MPC++ [10], OpenC++ [5], EC++ [4] or JTRANS [12] have solved this problem. These systems allow the implementation of language extensions as modules that can be selected by users. However, extensibility of syntax is slightly restricted in these systems.

2.2 Description of EPP

The Extensible Java PreProcessor kit, *EPP*, is an application framework for preprocessor-type language extensions. EPP is itself a source-to-source preprocessor of Java. The recursive-descent parser[1] of EPP provides many hooks for extensions (described in Sect. 2.4). By using these hooks, the extension programmer can introduce new features, possibly associated with new syntax, without editing the source code of EPP. Because all grammar rules are defined in a modular way, it is also possible to remove some original grammar rules from standard Java.

Once the parsed program has been transformed into a tree, a preprocessor programmer can easily manipulate it from a program. The usefulness of this kind of tool has already been proven by Lisp implementations, and has been adapted to C++ by various systems like Sage++ [2], MPC++ [10] and OpenC++ [5].

EPP enables preprocessor programmers to write an extension as a separate module. We call the extension modules *EPP plugins*. Several plugins can be programmed with EPP, then assembled together if they do not cause conflicts (among identifiers, for example). Users of the preprocessor can select any plugin that fits the characteristics of their projects.

Composability of EPP plugins can be realized thanks to a description language, *Ld-2* [9], an object-oriented package implemented in Common Lisp [15]. The inheritance mechanism of object-oriented languages makes it easy to implement extensible applications because all methods of objects can be considered as hooks for extensions. In addition to the traditional inheritance mechanism, Ld-2 provides a novel feature called *system mixin*, which supports extensible and flexible software.

Although the current target of EPP is only Java, the architecture of EPP is applicable to other programming languages.

2.3 An Example of an EPP Plugin

The user of EPP can specify one or more *EPP plugin files* at the start of a Java program. Figure 1 is a simple example program using an EPP plugin that defines a swap macro. The plugin file is actually an Ld-2 program file that will be dynamically loaded by EPP before preprocessing.

```
#epp load "swap"
public class test {
  public static void main(String[] argv){
    int a = 1, b = 2;
    swap(int, a, b);
  }
}
```

Fig. 1. A program using a swap plugin

2.4 System Mixin

The first program in Fig. 2 defines a method named :m which contains nested if statements. This method is written in a traditional, "monolithic" way. When a programmer wants to add a new else-if clause to this method, the only way to do so is to edit the source code. By using the inheritance mechanism, some editing can be avoided, but this changes the identity of the method. It is indeed impossible to extend the behavior of the method without defining a new subclass.

Ld-2 provides a novel mechanism, *system mixin*, which enables the program to be split into small reusable modules. Bracha[3] showed the flexibility and reusability of *mixin-based* inheritance. System mixins are similar to mixins, but the unit of inheritance is not a class but the whole program.

The second program in Fig. 2 uses the system mixin feature and has the same meaning as the first one. However, its structure is much more modular.

```
(class <foo> ()
  (defmethod :m (data)
    (if (eq data 'B)
        (do-B)
        (if (eq data 'A)
            (do-A)
            (do-default-behavior)))))
```

```
(defsystem <<skeleton>> ()
  (class <foo> ()
    (defmethod :m (data)
      (do-default-behavior))))
(defsystem <<A>> ()
  (class <foo> (data)
    (method :m (data)
      (if (eq data 'A) (do-A) [:original data]))))
(defsystem <<B>> ()
  (class <foo> (data)
    (method :m (data)
      (if (eq data 'B) (do-B) [:original data]))))
(setup (list <<B>> <<A>> <<skeleton>>))
```

Fig. 2. Modular definitions in Ld-2

A `defsystem` defines a system mixin, which is a program fragment. The `setup` statement in the last line specifies the system mixins that should be composed into the complete program. The expression [`:original data`] is similar to the method invocation to the super class in traditional object-oriented languages. When the method `:m` is called, the fragment of method `:m` defined in <<B>> will be called. If the fragment evaluates [`:original data`], the next fragment defined in <<A>> will be called.

The recursive-descent parser of EPP uses the system mixin feature. Each function which parses a non-terminal has hooks for extension that are actually methods which can be extended by other system mixins. The plugin programmer can add new `else-if` clauses to the methods to add new operators or new statements.

3 An Example of an EPP Plugin

In this section we introduce *Tiny Data-Parallel Java* as an example of an EPP plugin, show how it is translated, and discuss its performance.

3.1 Tiny Data-Parallel Java

As an example of EPP plugin, we implemented the *Tiny Data-parallel Java plugin* using the same translation technique as Data-Parallel C [7]. Some methods

```
#epp load "datap01"
public class CalcPi {
  public static void main(String argv[]){
    // Number of virtual processors.
    int num = 10000000;
    // Number of Java threads used for parallel execution.
    DpCalcPi.PE_n = 16;

    DpCalcPi.width = 1.0 / num;
    DpCalcPi p = new DpCalcPi(num);
    p.run();                // Execute the data-parallel method.
    System.out.println(DpCalcPi.result * DpCalcPi.width);
  }
}
parallel class DpCalcPi {
  static double result; // Mono variable.
  static double width;  // Mono variable.
  double x;             // Poly variable.
  public void run(){    // Data-parallel method.
    x = (VPID() + 0.5) * width;
    // Reduction. Calculate Σ 4 / (1 + x^2) .
    set_sum(result, 4.0 / (1.0 + x * x));
  }
}
```

Fig. 3. A program that calculates π.

of Java objects are treated as data-parallel methods. From the programmer's point of view, the methods are executed on a large number of virtual processors.

Figure 3 shows a program which calculates the value of π. The program launches ten million virtual processors at the same time, each calculating a part of the surface giving us the value of π. The run method of class DpCalcPi, which is defined with the special modifier parallel, is the data-parallel method which is executed in parallel by virtual processors. EPP translates this method into a standard Java program that uses thread libraries and synchronization primitives. Other methods are not changed by the translator.

In parallel classes, there are two type of variables, mono- and poly-variables. Mono-variables are shared by all virtual processors and poly-variables are owned by each virtual processor. Static variables defined in the parallel class are mono-variables, while instance variables are poly-variables. Mono- and poly-variables can be used in arbitrary expressions in the data-parallel method. It is possible to access a poly-variable on another virtual processor by using the expression x@[i].

The current implementation imposes some restrictions on how to write data-parallel methods. For example, no local variable can be declared within the data-parallel method, and remote variable accesses are not allowed in the middle of other expressions. In addition, current implementation performs no optimization.

However, we think that this extension is a good demonstration of the suitability of EPP to language extension, in particular in the field of parallel programming.

3.2 The Translator

The source code of Tiny Data-Parallel Java is expanded into plain Java code that contains either for statements (simulating virtual processors) or thread/process creation instructions, and synchronization. Figure 4 shows some source code using the virtual processor ID (VPID) and its translation into plain Java. Virtual

```
x = VPID();
System.out.println("VPID = " + x);
```

```
for (vpi = (0); (vpi) < (vpn); vpi++) {
  (vp_x)[vpi] = (((PE_i) * (vpn)) + (vpi));
  System.out.println(("VPID = ") + ((vp_x)[vpi]));
}
sync();
```

Fig. 4. Source code and translated code of a Tiny Data-Parallel Java program

processors can be allocated to several Java threads, PE_i being the ID of each thread; vpn gives the number of virtual processors allocated to threads. In reality, vpi is only the loop counter used to simulate the allocated virtual processors. Each poly-variable is represented as an array whose size is vpn. Therefore, a poly-variable x is translated to (vp_x)[vpi]. The method invocation sync() executes a barrier synchronization of all threads. This synchronization call is inserted in several places, such as before and after remote data access or reduction function calls.

3.3 Performance

Parallel execution of Tiny Data-Parallel Java has been tested in the following environments:

- A dual-processor Pentium Pro machine running Windows NT and JDK;
- A Cray CS6400 (a shared-memory machine with 32 SPARC processors) runing Solaris 2.5 and Toba, a Java-to-C translator which supports multi-threading.

Figure 5 shows execution time of the program which calculates π with 10 million virtual processors.

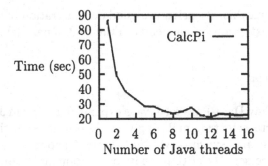

Fig. 5. Performance of π calculation on CS6400

4 Conclusion

We described the extensible preprocessor kit EPP, its description language Ld-2, and Tiny Data-Parallel Java as an example of EPP plugin. Ld-2 has a feature called system mixin which enables the program to be split into reusable modules. EPP is an application framework for preprocessors. The preprocessor programmer can implement EPP plugins, which are reusable and composable extension modules. EPP plugins can add new grammar rules to Java. Users can easily select the EPP plugins they need.

EPP and Ld-2 are currently implemented in Common Lisp, but we are implementing an EPP plugin which adds system mixin features to Java. EPP may then be rewritten in Java with this plugin.

References

1. A.V. Aho, R. Sethi, and J.D. Ullmann. *Compilers: Principles, Techniques and Tools*. Addison-Wesley Publishing company, 1987.
2. F. Bodin et al. Sage++: A Class Library for Building Fortran and C++ Restructuring Tools. In *Proc. of Object-Oriented Numerics Confs., Oregon*, April 1994.
3. G. Bracha and W. Cook. Mixin-based Inheritance. In *Proc. of ECOOP/OOPSLA '90*, 1990.
4. D. Caromel et al. Europa Parallel C++ version 2.1. In *Future Directions for Parallel C++*, June 1997. www.etl.go.jp/etl/bunsan/~roudier/postscripts/europa.ps.gz.
5. S. Chiba. A Metaobject Protocol for C++. In *Proceedings of OOPSLA '95*, volume 30(10) of *ACM Sigplan Notices*, pages 285–299, Austin, Texas, October 1995. ACM Press.
6. J. Gosling, B. Joy, and Steele. G. *The Java Language Specification*. Java Series. Sun Microsystems, 1996.
7. P.J. Hatcher and M.J. Quinn. *Data-Parallel Programming on MIMD Computers*. MIT Press, 1991.
8. Y. Ichisugi. EPP and Lods home page. www.etl.go.jp/etl/bunsan/~ichisugi.
9. Y. Ichisugi. Ld-2 users manual (DRAFT). Technical report, Electrotechnical Laboratory, 1997. can be obtained from [8]. In Japanese.

10. Y. Ishikawa. Meta-level Architecture for Extendable C++, Draft Document. Technical Report TR-94024, Real World Computing Partnership, 1994.
11. Gregor Kiczales, Jim des Rivières, and Dan Bobrow. *The Art of the Meta Object Protocol.* MIT Press, 1991.
12. A. Kumeta and M. Komuro. Meta-Programming Framework for Java. In *The 12th workshop of object oriented computing WOOC'97, Japan Society of Software Science and Technology*, March 1997.
13. P. Launay and J.L. Pazat. Integration of control and data parallelism in an object oriented language. In *Sixth Workshop on Compilers for Parallel Computers (CPC'96) Aachen, Germany*, December 1996.
14. NPAC. HPJava reference and workshops site. www.npac.syr.edu/projects/javaforcse/.
15. G.L. Steele. *Common Lisp the Language, 2nd edition.* Digital Press, 1990.

Numerical Solution of PDEs on Parallel Computers Utilizing Sequential Simulators

Are Magnus Bruaset[1], Xing Cai[2], Hans Petter Langtangen[2], and Aslak Tveito[2]

[1] Numerical Objects AS
[2] University of Oslo

Abstract. We propose a strategy, based on domain decomposition methods, for parallelizing existing sequential simulators for solving partial differential equations. Using an object-oriented programming framework, high-level parallelizations can be done in an efficient and systematical way. Concrete case studies, including numerical experiments, are provided to further illustrate this parallelization strategy.

1 Introduction

The purpose of this paper is to address the following problem: How can existing sequential simulators be utilized in developing parallel simulators for the numerical solution of partial differential equations (PDEs)? We discuss this issue in the light of object-oriented (O-O) programming techniques, which can substantially increase the efficiency of implementing parallel PDE software. Our parallelization approach is to use domain decomposition (DD) as an overall numerical strategy. More precisely, DD is applied at the level of subdomain simulators, instead of at the level of linear algebra. This gives rise to a simulator-parallel programming model that allows easy migration of sequential simulators to multiprocessor platforms. We propose a generic programming framework in which the simulator-parallel model can be realized in a flexible and efficient way. The computational efficiency of the resulting parallel simulator depends strongly on the efficiency of the original sequential simulators, and can be enhanced by the numerical efficiency of the underlying DD structure. This parallelization strategy enables a high-level parallelization of existing O-O codes. We show that the strategy is flexible and efficient and we illustrate the parallelization process by developing some concrete parallel simulators in Diffpack [DP].

2 Domain Decomposition and Parallelization

Roughly speaking, DD algorithms[SBG96] search the solution space of the original large problem by iteratively solving many smaller problems over subdomains. They are very efficient numerical techniques for solving large linear systems, even on sequential computers. Such methods are attractive for parallel computing if the subproblems can be solved concurrently.

In this paper we concentrate on a particular DD method called the (overlapping) *additive Schwarz method* (see [SBG96]). In this method the subdomains form an overlapping covering of the original domain. The method can be formulated as an iterative process, where in each iteration we solve updated boundary value problems over the subdomains. The work on each subdomain in each iteration consists mainly of solving the PDE(s) restricted to a subdomain using values from its neighboring subdomains, computed in the previous iteration, as Dirichlet boundary conditions. The subproblems can be solved in parallel because neighboring subproblems are only coupled through previously computed values in the overlapping region between the non-physical inner boundaries. The convergence of the above iterative process depends on the amount of overlapping. It can be shown for many elliptic PDE problems that the additive Schwarz method has an optimal convergence behavior, for a fixed level of overlapping, when an additional coarse grid correction is applied in each iteration (see [CM94] for the details).

3   A Simulator-Parallel Model in Diffpack

We propose a *simulator-parallel* model for parallelizing existing sequential PDE simulators. The programming model uses DD at the level of subdomain simulators and assigns a sequential simulator to each processor of a parallel computer. The global administration and communication related to parallel computation can be easily realized in an O-O framework. We have tested this parallelization approach within Diffpack [DP], which is an O-O environment for scientific computing with particular emphasis on numerical solution of PDEs.

The C++ Diffpack libraries contain user-friendly objects for I/O, GUIs, arrays, linear systems and solvers, grids, scalar and vector fields, visualization and grid generation interfaces. The flexible and modular design of Diffpack allows easy incorporation of new numerical methods into its framework, whose content grows constantly. In Diffpack, a finite element (FE) based PDE solver is typically realized as a C++ class having a grid, FE fields for the unknowns, the integrands in the weak formulation of the PDE, a linear system toolbox, some standard functions for prescribing essential boundary conditions, and some interface functions for data input and output. The above parallelization approach offers flexibility in the sense that different types of grids, linear system solvers, preconditioners, and convergence monitors are allowed for different subproblems. A new parallel simulator can thus be derived from reliable sequential simulators in a flexible and efficient way.

Diffpack was originally designed without paying particular attention to parallel computing. A large number of flexible, efficient, extensible, reliable, and optimized sequential PDE solver classes have been developed. The parallelization strategy presented in this paper extends Diffpack by offering the means to adapt these sequential solvers for concurrent computers. It is hoped that this can be done without loss of computational efficiency as compared to a special-purpose application code which is implemented particularly for parallel computing. Nor-

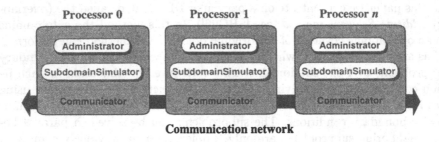

Fig. 1. A generic framework for the simulator-parallel programming model.

mally, such special-purpose parallel codes employ *distributed* arrays, grids and so on. That is, the conceptual model contains abstractions representing *global* quantities. In our simulator-parallel model proposed above, we only work with local PDE problems. This avoids the need for data distribution in the traditional sense. We are only concerned with a standard, sequential PDE solver on each processor and some glue for communicating boundary values for the local problems, in addition to an overall numerical iteration. O-O programming is a key ingredient that makes this migration to parallel computers fast and reliable. The advantages of such an approach are obvious and significant: (1) optimized and reliable existing sequential solvers can be re-used for each subproblem, (2) message passing statements are kept to a minimum and can be hidden from the application programmer in generic objects, and (3) the extra glue for communication and iteration is just a short code at the same abstraction level as the theoretical description of the parallel DD algorithm. We believe that the suggested parallel extension of Diffpack will make it much easier for application developers to utilize parallel computing environments. The fundamental question of the efficiency of our general high-level approach will be addressed in the numerical experiments.

4   A Generic Programming Framework

To increase flexibility and portability, the simulator-parallel model is realized in a generic programming framework consisting of three main parts (see Fig. 1): a SubdomainSimulator, a Communicator, and an Administrator. Furthermore, each of the three parts of the programming framework is implemented as a C++ class hierarchy, where different subclasses specialize in different types of PDEs and specific numerical methods. In this way a programmer can quickly adapt his existing sequential simulator(s) into the framework. The subsequent discussion requires some knowledge of O-O programming techniques.

The base class SubdomainSimulator is a generic interface class offering an abstract representation of a sequential simulator. First, the class contains *virtual* functions that are to be overridden in a derived class to make the connection between SubdomainSimulator and an existing sequential simulator. Second, the class contains functions for accessing the subdomain local data needed for communication among neighboring subdomains. We have introduced subclasses of SubdomainSimulator to generalize different specific simulators. As examples, SubdomainFEMSolver is designed for simulators solving a scalar/vector elliptic PDE discretized by FE methods, and SubdomainFDMSolver is for simulators using finite difference discretizations.

The base class Communicator is designed to generalize the communication between neighboring subdomains. The primary task of the class is to determine the communication pattern by finding where the neighboring subdomain simulators are located, and which part of the local data should be sent to each neighbor. A hierarchy of different communicators is also built to handle different situations. One example is CommunicatorFEMSP which specializes in communication between subdomain simulators using the FE discretization. More importantly, the concrete message passing model, which is MPI in the current implementation, is hidden from the user so that a new message passing model may be easily inserted without changing the other parts of the framework.

Finally, Administrator performs the numerics underlying the iterative process of DD. Coordinating with each other, an object of Administrator exists on each subdomain. Typically, Administrator has a SubdomainSimulator and a Communicator under its control, so that SubdomainSimulator's member functions are invoked for computation and member functions of Communicator are invoked for communication. A hierarchy of administrator classes is built to realize different solution processes for different model problems. For example, class PdeFemAdmSP handles cases of a scalar/vector elliptic PDE discretized by FE methods.

5   Some Numerical Experiments

In this section, we apply the proposed parallelization strategy to develop parallel simulators for solving the Poisson equation, the linear elasticity problem and the two-phase porous media flow problem, respectively. The implementation of the parallel simulators consists essentially of extending the existing Diffpack sequential simulators to fit in the programming framework described above. Since we apply the FE discretization for elliptic PDEs, the situation for the first two test problems is straightforward, because only a single elliptic PDE is involved in each case, even though it is a scalar equation in one case and a vector equation in the other. The classes CommunicatorFEMSP and PdeFemAdmSP can be used directly. So the only necessary implementation is to derive a new simulator class as a subclass of both the existing sequential simulator and class SubdomainFEMSolver. Using a subclass for gluing the sequential solver and the parallel computing environment avoids *any modifications* of the original sequential solver class. The parallelization was done within an hour. The situation for the third test problem is slightly more demanding because the system consists of an elliptic PDE and a hyperbolic PDE. The two PDEs are solved with different numerical methods, which means additional work for implementing a suitable administrator as

Table 1. CPU consumption (in seconds) of the iterative DD solution for different M. (The fixed total number of unknowns is 481×481.)

P	M	CPU	Speedup	Subdomain grid
1	4	44.41	—	257×257
4	4	11.37	3.91	257×257
6	6	7.92	5.61	257×181
8	8	5.84	7.60	257×135
12	12	3.92	11.33	257×91
16	16	3.06	14.51	257×69

a subclass of `PdeFemAdmSP`. However, programming in the proposed framework enables a quick implementation which was done in a couple of days.

In the following, we give for each test problem the mathematical model and CPU consumption measurements associated with different numbers of processors in use. The measurements are obtained on a SGI Cray Origin 2000 parallel computer with R10000 processors. The resulting parallel simulators demonstrate not only nice scalability, but also in some simulations the intrinsic numerical efficiency of the underlying DD algorithm.

5.1 The Poisson Equation

Our first test problem is the 2D Poisson equation on the unit square, with a right hand side and Dirichlet boundary conditions so that $u(x, y) = -xe^y$ is the solution of the problem.

We denote the number of subdomains by M and introduce an underlying uniform 481×481 global grid covering Ω. (The grid is only used as the basis for partitioning Ω for different choices of M.) With the number of processors P equal to M, we study the CPU consumption of the DD solution process with different choices of P. Suitable coarse grid corrections, together with a fixed level of overlapping, are used to achieve the convergence up to a prescribed accuracy under the same number of DD iterations, independently of P. The largest coarse grid problem is considerately smaller than any subproblem, so the CPU time spent on the coarse grid correction is negligible. To obtain a reference CPU consumption associated with a single processor, we make a sequential implementation of the same iterative DD process and measure the CPU consumption with $M = 4$. In this test problem we are able to use a fast direct solver for the subproblems. The important property is that its computational work is linearly proportional to the number of unknowns. From the numerical experiments, we observe that the communication overhead is very small in comparison with the solution of the subproblems. These two factors together contribute to the speedups in Table 1.

Although DD methods achieve convergence independently of the problem size, it may be more efficient to use a standard Krylov subspace method for this problem when only a small number of processors are available. So this test

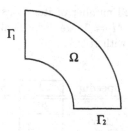

Fig. 2. The solution domain for the linear elasticity problem.

problem mainly verifies that our implementation of the generic programming framework has kept the communication overhead at a negligible level. In the following case (solving a linear elasticity problem) we will see that the parallel DD approach shows its numerical efficiency even for a smaller discrete problem.

5.2 The Linear Elasticity Problem

The 2D deformation of an elastic continuum subject to a plain strain can be modelled by the following partial differential equation

$$-\mu\Delta\mathbf{U} - (\mu + \lambda)\nabla\nabla \cdot \mathbf{U} = \mathbf{f},$$

where the displacement field $\mathbf{U} = (u_1, u_2)$ is the primary unknown and $\mathbf{f}$ is a given vector function. Here μ and λ are constants. The 2D domain is a quarter of a hollow disk (see Fig. 2). On the boundary the stress vector is prescribed, except on Γ_1, where $u_1 = 0$ and on Γ_2, where $u_2 = 0$.

We use a structured 241×241 curvilinear global grid as the basis for partitioning Ω into M overlapping subdomain grids. Again we have $P = M$. In this test problem we apply an inexact solver for the subproblems. More precisely, a conjugate gradient method preconditioned with incomplete LU-factorization (ILU) is used to solve the subproblems. Convergence of the subproblems is considered reached when the residual of the local equation system, in the discrete L_2-norm, is reduced by a factor of 10^2. For the global convergence of the DD method we require that the global residual, in the discrete L_2-norm, is reduced by a factor of 10^4. In Table 2, we list the CPU consumptions and the number of iterations I used in connection with different M. The fixed number of degrees of freedom is 116,162. To demonstrate the superior efficiency of the parallel DD approach, we also list the CPU consumption and number of iterations used by a sequential conjugate gradient method, preconditioned with ILU, for the same global convergence requirement and problem size. Note that the super linearity of the speedup is due to the fact that the preconditioned conjugate method works more efficiently for smaller subproblems.

Table 2. CPU consumptions and number of iterations used by the parallel DD approach. (The measurements for $M = 1$ are associated with a sequential conjugate gradient method preconditioned with ILU.)

M	CPU	Speedup	I	Subdomain grid
1	110.34	–	204	241×241
2	40.97	2.69	7	129×241
4	23.20	4.76	11	129×129
6	12.43	8.88	8	93×129
8	7.95	13.88	7	69×129
12	4.95	22.29	7	47×129
16	3.64	30.31	7	35×129

Table 3. Simulations of two-phase flow by a parallel DD approach. (Columns of total CPU consumptions are listed for 1. The whole simulation, 2. The pressure equation, 3. The saturation equation, accumulated from 744 discrete time levels.)

M	Total CPU	Subdomain grid	CPU pre. eq.	CPU sat. eq.
2	12409.01	129×241	11807.60	241.41
4	5343.50	129×129	5202.79	140.71
6	3201.89	129×91	3101.38	100.09
8	2827.77	129×69	2770.05	77.72
12	1881.33	91×69	1826.82	54.51
16	1388.75	69×69	1346.36	42.39

5.3 The Two-Phase Porous Media Flow Problem

We consider a simple model of two-phase (oil and water) porous media flow in oil reservoir simulation,

$$s_t + v \cdot \nabla(f(s)) = 0 \quad \text{in } \Omega \times (0, T], \tag{1}$$
$$-\nabla \cdot (\lambda(s)\nabla p) = q \quad \text{in } \Omega \times (0, T]. \tag{2}$$

In the above system of PDEs, s and p are the primary unknowns and $v = -\lambda(s)\nabla p$. We carry out simulations for the time interval $0 < t \leq 0.4$ seconds. At each discrete time level, the two PDEs are solved in sequence. The hyperbolic Equation (1), referred to as the saturation equation, is solved by an explicit finite difference scheme, whereas the elliptic Equation (2), referred to as the pressure equation, is solved by a FE method. The 2D spatial domain is the unit square with impermeable boundaries. An injection well (modelled by a delta function) is located at the origin and a production well at (1,1). Initially, $s = 0$ except at the injection well, where $s = 1$.

A uniform 241×241 global grid is used to cover Ω. For all the discrete time levels, the convergence requirement for the DD method when solving (2) is that the absolute value of the discrete L_2-norm of the global residual becomes smaller than 10^{-8}. We have used a conjugate gradient method preconditioned with ILU as the inexact subdomain solver, where local convergence is considered reached when the local residual in the discrete L_2-norm is reduced by a factor of 10^3. Results are shown in Table 3.

Acknowledgments This work has received support from The Research Council of Norway (Programme for Supercomputing) through a grant of computing time. The authors also thank Dr. Klas Samuelsson for numerous discussions on various subjects.

References

[CM94]   T.F. Chan and T.P. Mathew. Domain decomposition algorithms. *Acta Numerica*, pages 61–143, 1994.
[DP]   Diffpack Home Page. http://www.nobjects.com/Diffpack.
[SBG96] B.F. Smith, P.E. Bjørstad and W.D. Gropp. *Domain Decomposition, Parallel Multilevel Methods for Elliptic Partial Differential Equations*. Cambridge University Press, 1996.

The Trio-Unitaire Project:
A Parallel CFD 3-Dimensional Code

C. Calvin[1] and Ph. Emonot[1]

CEA - Grenoble, DRN/DTP/SMTH
17, rue des Martyrs - 38054 Grenoble cedex 09 - France

Abstract. The structure of a new generation of thermalhydraulic code: Trio-Unitaire is presented in this paper. This code has been designed to solve large 3D structured or unstructured CFD problems. The solutions adopted to achieve this goal (object-oriented design and parallelism) are described and the paper focuses on the technical solutions used. Some preliminary experimental results on a Cray T3E are presented.

1 INTRODUCTION

1.1 Main objectives

Trio Unitaire (TRIO-U) is a project for the development of a parallel object-oriented thermo-hydraulic 3D code. The three main goals of the project are as follows:

1. The solving of huge 3D structured and unstructured problems: We want to model very thin physical phenomena, like turbulent flow in large complex geometry.
2. This project has to be an open structure for the development, coupling and integration of other applications: integration of different physical models (compressible, incompressible, turbulence, ...), development and integration of new numerical methods and discretization (finite element, finite volume, finite volume elements, iterative solvers, multigrid methods, etc.) and coupling between various physical equations (thermal, hydraulic, mechanic,...).
3. The portability: the code may be run on many different computers: workstations, parallel shared memory machines, parallel distributed memory machines.

In order to achieve the first objective the code is intrinsically parallel. The simulation of thin 3D physical phenomena requires meshing with several million elements, so that the use of parallel distributed memory machines is compulsory. But, the code could also run on other architectures, like parallel shared memory machines, networked workstations, or even in a sequential way on a single workstation. So the structure of the code has to take into account all these parameters.

In order to achieve the portability and open structure objectives, an object-oriented conception, OMT [16] is used. The implementation language chosen is C++ [18].Thus, all the parallelism management and communication routines have been encapsulated. We have defined parallel I/O and communication classes over standard I/O streams of C++, which allow the developer an easy use of the different modules of the application without dealing with basic parallel process management and communications. Moreover, the encapsulation of the communication routines guarantees the portability of the application and allows efficient tuning of basic communication methods in order to achieve the best performance on the target architecture. At the present time, the encapsulated message passing library are PVM [1] or MPI [20].

Many people and laboratories are involved in the development of parallel CFD codes. Most of these projects are preliminary study, concerning only some basic kernels of a CFD code. The aim of these kind of studies is either to achieve some benchmarks [8, 2] or to develop a prototype of an industrial parallel code. Some of them are the parallelization of existant sequential industrial codes like N3S [3, 4]. In this case, each new development of the sequential code implies some new efforts to parallelize them [11]. The last category of projects concerns the design of tools and libraries for the development of parallel CFD applications [14].
Our approach is slightly different: we develop a parallel industrial code, and the new physical and numerical methods introduced in the code do not lead to new developments for their parallelization.

1.2 Organization of the paper

We describe, in this paper, the structure of the parallel code and the main concepts which have driven the design. Although, the code is still under development and validation, we present some first performances results obtained on a Cray T3E.
In the first section, the different models of parallelism used for the conception of the code are presented. In a second part, the parallel structure of the application and the different classes designed to manage the parallelism and the communication are described. The domain decomposition method used and the parallelization of the solver are discussed. Finally, some first experimental results of a 3D simulation on Cray T3E are presented.

2 PARALLELISM MODELS

The size of the considered problems can not be solved on standard machines. Moreover, even for the problems that can be solved on such machines, CPU times are prohibitive. So, we have to consider massiveley parallel computers with distributed memory in order to solve large thin 3D problems. However, to achieve the goals of generality of the code, we have also to consider standard machines, like sequential or shared memory computers. One of the originality of the project is to create an intrinsically parallel application. The code is designed in order to hide the parallelism inside the structure. This allows the development of new modules without taking care about their parallelization. Moreover, the same code can be run either on a parallel machine or a sequential one without any change, even in the input data.
The following designing choices have been done:

– **Parallelization model :** Data parallelism [7]. The initial domain is split into smaller ones and each of these sub-domains are distributed among the available processors.

We use a domain decomposition method which is quite different from the standard ones, as presented in a subsequent section.

- **Programming model:** SPMD [7]. The same code is executed by all the processors but using different data.
- **Communication model:** Message passing [7]. Since we use irregular data structures, the communications between processors are explicit to achieve good performance. At the present time, the library used is PVM [1], but the use of other message passing libraries, like MPI [20], can be done very quickly.

3 PARALLEL CODE ARCHITECTURE

As we have mentioned ealier, a single version of the code has been developed, which can be executed on both sequential and parallel machines. We consider that a sequential execution is a parallel one using only one processor. This implies some constraints on the overall design of code.

3.1 Parallelism management classes

We have designed a model of a parallel code: the class Process is the model of a computational process. Each object of the hierarchy inherits from the class Process (see figure 1). Thus, each object can communicate with the other ones in a natural way.

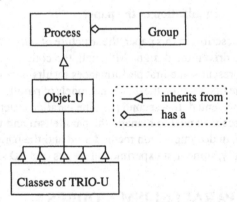

Fig. 1. Hierarchy of the classes

3.2 Communication and parallel Input/Output methods

The main idea which has driven the design of these classes, is transparency: a message is sent to another processor in the same way as data is written in a file.

Figure 3 illustrates the use of the communication classes. This program exchanges data between two processors.

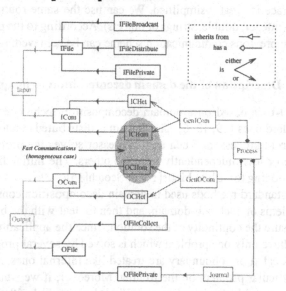

Fig. 2. Hierarchy of the communication and parallel I/Oclasses.

```
{// Program executed by all the proc.
    double array[10];
    char string[80];
    ArrOfInt X;        // Class of the hierarchy
    SCHom SendBuffer;
    ECHom ReceiveBuffer;

    // Proc. 0 sends the message array to the proc. 1,
    SendBuffer << x << string << array << flush(0,1,99);

    // Proc. 1 receives the message from the proc. 0.
    ReceiveBuffer(0,1,99) >> x >> string >> array;
}
```

Fig. 3. Example of communication classes usage.

The same principle is used for the classes which redefine the I/O. According to the file type, the behaviour is different, although the code expression is identical.

The communication methods are based on these previous classes. They implement the main communication schemes [12, 9]:

- Point-to-point communications : an exchange of data between two processors.
- Broadcasting : the same information is sent to all the processors.
- Distribution : different data is sent to each processor.
- Synchronisation.

Due to the polymorphism property of C++, there are only 4 routines. Thus the user interface is greatly simplified. We can use the same routines for sending any kind of data (integer, double, strings or objects). According to the parameters of the routine any of the previous communication scheme can be achieved.

3.3 Description of the domain decomposition technique used

TRIO-U is based on a domain decomposition technique which is different from the standard ones [19, 13, 6]. The domain is distributed in a load-balanced way among the different processors. Then each processor solves the problem on its own sub-domain more or less independently from the others. The initial distribution is achieved using partitioning tools like Metis[10] or Scotch[15].

The standard methods used in domain decomposition consist in solving in parallel the problems of each sub-domain, and then to deal with the boundaries problems in order to insure the continuity of the solution. Since the application is intrinsically parallel, we still have only one problem which is solved by several processors. The elements which are located on a boundary are treated like internal ones, and we do not have to solve a particular problem on the frontier. Moreover, if we want to develop new numerical methods or physical models, we will not have to deal with the parallelism, since the data structure is parallel. To achieve this, the objects which represent the geometry and its discretization carry some information about the interface between sub-domains. So the connectivity between sub-domains is contained in the data structure, and each object of the geometry thus knows how to get information of a next sub-domain. In order to optimize communications, we have implemented an overlap sub-domain method, so the values of boundaries are only exchanged at each time step of the simulation.

We have defined the notion of distributed arrays. Each distributed array owns a personal part which contains the values of its sub-domain and a virtual part which contains a copy of the values of the next sub-domains.

We present in figure 4 an example of the internal representation of the pressure values for one processor onto a domain decomposed into 3 sub-domains. We have represented in this example, the personnal part and the virtual part of the pressure vector hold by processor 0, and the informations concerning its neighbors.

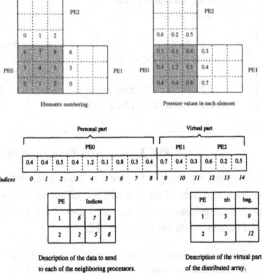

Fig. 4. Example of a distributed array which represents the elements list of a 3x3 sub-domain with 2 neighbours.

3.4 Resolution of equation systems

The algorithms to solve the equation systems are based on the conjugate gradient algorithms [5, 17]. The basic numerical kernels which are involved in these kinds of algorithms are:

- Matrix-Vector product,
- Scalar product,
- Norm computation for the stopping criterion.

Since a matrix object inherits from a distributed array object, it can provide the parallel version of the previous numerical kernels. Thus, we can implement new solvers based on the conjugate gradient algorithm which will be automatically parallel.

We have already implemented three different conjugate-gradient-based algorithms with different pre-conditioners. They have been designed in such a way, that they require no extra-communications.

We have also planned the use of parallel numerical libraries which provide efficient parallel pre-condioners and solvers. The encapsulation of these kinds of libraries can be easily done using the object structure of TRIO-U.

4 FIRST EXPERIMENTAL RESULTS

We present now some experimental results on a simple thermal-hydraulic 3D problem. The initial domain is composed of 576,000 elements meshing and the same problem has been solved on a Cray T3E using three different kinds of partition: slices along X axis, slices along Z axis and blocks. Figure 5 represents the execution times obtained.

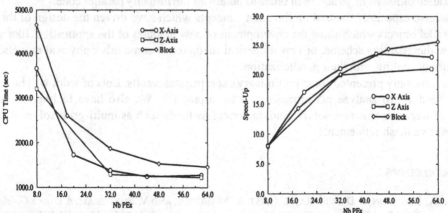

Fig. 5. Execution times on T3E for 576,000 elements.

Fig. 6. Speed-ups on T3E for 576,000 elements.

As one can observe, we obtain a good decreasing of execution times. We can also see a great influence of the partition method used. These first observations are confirmed by the curves in figure 6.

The performances obtained are quite good since we obtain a speed-up of 20 for 32 processors. The efficiency is still at about 55% on 40 processors, although the problem is quite small for a single processor.

We can notice that even if the block partitioning, is less efficient for a small number of processor, the speed-up is better than the other methods. This kind of behavior is explained by the number of iterations of the conjuguate gradient solver which is greater for a block partitioning, as one can see on figure 7.

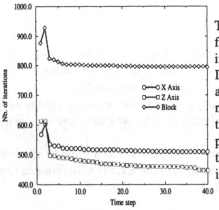

This effect can be explained mostly by the fact that the matrix conditioning is greatly influenced by the partitioning method used. Indeed, since we use a block-ssor method as a pre-conditioner, the terms which are neglected are not the same depending on the initial domain splitting. It seems that a partitioning which is orthogonal or parallel to the flow is better for the marix conditioning.

Fig. 7. Number of iteration of the conjuguate gradient on 32 processors.

5 CONCLUSION

We have presented in this paper the architecture and design of a new generation CFD code. The main originalities of this code are its object oriented conception and the parallelization of its structure in order to obtain an intrinsically parallel code.

We have especially focused on the main concepts which have driven the design of the parallel objects which allow the development of new modules of the application, like a new dicretization scheme, or new numerical methods, or some other physical models, without dealing with their parallelization.

We have only presented some preliminary experimental results. Lots of work will have to be done to analyze performance and to improve it. We also have to implement and integrate some new solvers and numerical methods such as multi-grids solver and adaptive mesh refinements.

References

1. A. BEGUELIN, J. DONGARRA, A. GEIST, R. MANCHEK, AND V. SUNDERAM, *A User's Guide to PVM Parallel Virtual Machine (version 3)*, tech. rep., Oak Ridge National Laboratory, May 1994.
2. C. CRAWFORD, D. NEWMAN, AND G. KARNIADIS, *Parallel Benchmarks of Turbulence in Complex Geometries*, in Proceedings of Parallel CFD'95, 1995.
3. G. DEGREZ, L. GIRAUD, M. LORIOT, A. MICELOTTA, B. NITROSSO, AND A. STOESSEL, *Parallel Industrial CFD Calculations with N3S*, in Proceedings of HPCN'95, Lecture Notes in Computer Science, 1995, pp. 820–825.

4. C. FARHAT AND S. LANTERI, *Simulation of Compressible Viscous Flows on a Variety of MPPs: Computational Algorithms for Unstructured Dynamic Meshes and Performances Results*, Comp. Meth. in Appl. Mech. and Eng., (1994), pp. 35–60.

5. G. GOLUB AND C. V. LOAN, *Matrix Computations*, The Johns Hopkins University Press, Baltimore, second edition ed., 1989.

6. K. HOFFMANN AND J. ZOU, *Parallel Efficiency of Domain Decomposition Methods*, Parallel Computing, (1993), pp. 1375–1391.

7. K. HWANG, *Advanced Computer Architecture - Parallelism, Scalability, Programmability*, Mc Graw-Hill and MIT Press, 1993.

8. D. JAYASIMHA, M. HAYDER, AND S. PILLAY, *Parallelizing Navier-Stokes Computations on a Variety of Architectural Platforms*, in Proceedings of Supercomputing'95, 1995.

9. JEAN DE RUMEUR, *Communications dans les Réseaux de Processeurs*, Masson, 1994.

10. G. KARYPIS AND V. KUMAR, *METIS: Unstructred Graph Partionning and Sparse Matrix Ordering. Version 2.0*, tech. rep., University of Minnesota, Dept. of Computer Science, Minneapolis, MN 55455, 1995.

11. S. LANTERI, *Parallel Solutions of Three-Dimensional Compressible Flows*, tech. rep., INRIA, Sophia Antipolis, 2004 Route des Lucioles, BP 93, 06902 Sophia-Antipolis, June 1995.

12. F. T. LEIGHTON, *Introduction to Parallel Algorithms and Architectures: Arrays - Trees - Hypercubes*, Morgan Kaufman Publishers, Inc., 1992.

13. G. MEURANT, *Domain Decomposition Methods for P.D.E.s on Parallel Computers*, International Journal Supercomputer Applications, (1988), pp. 5–12.

14. P. OLSSON, J. RANTAKOKKA, AND M. THUNÉ, *Software Tools for Parallel CFD on Composite Grids*, in Proceedings of Parallel CFD'95, 1995.

15. F. PELLEGRINI, *SCTOCH 3.1: User's Guide*, tech. rep., LaBRI, Université Bordeaux I, 351 cours de la libération, 33045 Talence, France, 1996.

16. J. RUMBAUGH, M. BLAHA, W. PREMERLANI, F. EDDY, AND W. LORENSEN, *Object Oriented Modeling and Design*, Prentice Hall, 1991.

17. Y. SAAD, *Iterative Methods for Sparse Linear Systems*, PWS Publishing company, 1996.

18. B. STROUSTRUP, *The C++ Programming Language. 2nd edition*, Addison Wesley Publishers, 1992.

19. P. L. TALLEC, *Domain decomposition methods in computational mechanics*, Computational mechanics advances, 1 (1994), pp. 121–220.

20. THE MPI FORUM, *Document for a Standard Message-Passing Interface*, Apr. 1994.

Overture: An Object-Oriented Framework for Solving Partial Differential Equations

David L. Brown, William D. Henshaw, and Daniel J. Quinlan

Los Alamos National Laboratory, CIC-19 MS B256, Los Alamos NM 87545, USA

Abstract. The **Overture** framework is a collection of C++ classes that can be used to solve partial differential equations (PDEs). These classes were designed to support applications in one, two and three space dimensions on geometries ranging from simple rectangular regions to complicated three dimensional domains, and to support adaptive mesh refinement and moving grids. **Overture** is designed to run on serial and parallel machines through the use of the A++/P++ serial/parallel array class library. The **Overture** classes hide the details of the underlying data-structures and hide the details of features common to many PDE solvers such as the implementation of finite-difference and finite-volume operators and boundary conditions. In addition to the Mapping and Grid classes that represent geometry, the GridFunction classes that represent solutions, and the operator classes, there are classes for high-level interactive plotting, and data-base management. **Overture** also includes sophisticated grid generation capabilities for creating overlapping grids.

1 Introduction

Overture is an object-oriented environment for solving partial differential equations on serial and parallel architectures. **Overture** contains a wide variety of classes that can be used to write sophisticated PDE solvers, to easily compute finite-difference and finite volume discretizations on curvilinear grids, to solve sparse matrix problems, to plot results, to save results, to generate grids, [1], [2]. The design of **Overture** has evolved over the past 15 years or so from the Fortran based CMPGRD environment to the current C++ version. Although the Fortran implementation was used for complicated three dimensional adaptive and moving grid computations, the programs were difficult to write and maintain. **Overture** was designed to have at least all the functionality of the Fortran code but to be as easy as possible to use; indeed, an entire PDE solver on an overlapping grid can be written on a single page as shown later in this paper.

A composite overlapping grid consists of a set of logically rectangular curvilinear computational grids that overlap where they meet and together are used to describe a computational region of arbitrary complexity. This method has been used successfully over the last decade and a half, primarily to solve problems involving fluid flow in complex, often dynamically moving, geometries. There

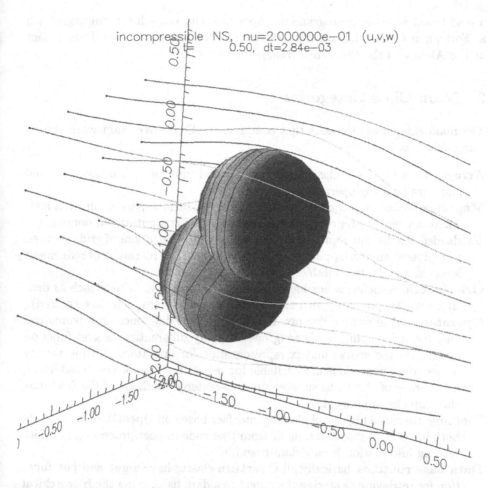

Fig. 1. Incompressible Navier-Stokes flow around two spheres, computed and plotted with Overture. The pressure is plotted on the surface of the spheres.

are a number of reasons why the design and implementation of a flexible over-lapping grid solver is quite complex. First of all the data structures required to hold geometry information (vertices, Jacobian, boundary normals, interpolation data, etc.) and the information for moving grids, adaptive grids and multigrid are extensive. Secondly, complicated partial differential equations are solved us-ing sophisticated numerical algorithms; a current application at Los Alamos involves low Mach-number combustion with many reacting species. In addition, techniques such as block structured AMR may be used to locally increase com-putational resolution and increase overall computational efficiency. If the simu-lation is to run on a parallel architecture, there are correspondingly more issues involved in writing the code. The net result of the data structures, advanced algorithms, and modern architectures is a PDE solver code that is an extremely complex system. Successfully writing, debugging, modifying and maintaining software that implements this system is a daunting if not impossible task using a traditional structured programming approach and procedural languages such as Fortran or C. This has provided the primary motivation for the development at Los Alamos of the **Overture** framework in C++.

2 Main Class Categories

The main categories of classes that comprise the **Overture** framework can be summarized as follows:

Arrays the A++/P++ class library defines multidimensional arrays for serial and parallel array operations[4].

Mappings define transformations such as curves, surfaces, areas, and volumes. These are used to represent the geometry of the computational domain.

Grids define a discrete representation of a mapping. Collections of grids are used to represent an overlapping grid and to represent the hierarchy of refinement levels in an adaptive mesh.

Grid functions define a flexible representation of field variables (such as density, velocity, pressure) with arbitrary centering (vertex, cell, face centred).

Operators define finite-difference and finite volume difference approximations on curvilinear domains allowing one to easily differentiate a grid function or to form the sparse matrix representation for operators. A wide variety of boundary conditions are available for use with PDE solvers. In addition, a hierarchy of classes to support upwind-centered methods of the Godunov class are also supported.

Plotting there is a high-level plotting interface based on OpenGL that supports both interactive graphics callable from user code to post-processing graphics reading information from a data-base file.

Data base routines basically all **Overture** classes have a **get** and **put** function for retrieving or storing the object in a data base, using the Hierarchical Data Format from NCSA.

AMR++ defines the operations common to block-structured adaptive mesh refinement (AMR) algorithms, and for patched-based domain decomposition methods.

Grid Generators a variety of techniques are available for generating curvilinear grids and a fully automatic algorithm for generating overlapping grids is provided.

3 A++ array class

The A++P++ array class library is extensively within **Overture** . It has been carefully described in a number of other documents, [4]. The basic functionality is perhaps best shown in a short example. Here is a program that uses A++ arrays to solve Poisson's equation.

```
// Solve u_xx + u_yy = f by a Jacobi Iteration
int n=100;
Range R(0,n);              // define a range of indices: 0,1,2..,n
```

```
floatArray u(R,R), f(R,R);      // declare two arrays
u=0.; f=1.;                     // initialize u, and the rhs, f
float h=1./n;                   // grid spacing
Range I(1,n-1),J(1,n-1);        // define ranges for the interior
for( int iteration=0; iteration<100; iteration++ )
{
  u(I,J)=.25*( u(I+1,J)+u(I-1,J)+u(I,J+1)+u(I,J-1) -f(I,J)*(h*h));
}
```

Notice how the Jacobi iteration for the entire array can be written in one statement. When linked with the P++ library this code will be a parallel code.

4 Mappings

Mappings are used to represent the geometry of the computational domain. Mappings can define lines, curves, surfaces, volumes, rotations, coordinates stretchings etc. Mappings contain a variety of information and functions that can be use useful for grid generators and solvers:

1. Mappings contain information about their domain space, range space, boundary conditions, singularities, and so on.
2. Mappings are easily composed allowing coordinate stretching, rotations, bodies of revolution, etc.
3. The *inverse* of the mapping is always defined (for a curve or surface the inverse is the closest point in the L_2 norm).
4. Since the Mapping still exists in a PDE solver, adaptive algorithms can access the original geometry to accurately determine the locations of refined points.

Here is an example where we create a mapping for an annulus and evaluate the transformation and its inverse. The annulus is a mapping from the unit square into cartesian space.

```
AnnulusMapping annulus;               // Create a mapping for an annulus
floatArray r(1,2),x(1,2),xr(1,2,2);   // arrays to evaluate mapping (1 point)
r=.5;                                 // r=(.5,.5)
annulus.map(r,x,xr);                  // evaluate the mapping and derivatives
annulus.inverseMap(x,r);              // evaluate the inverse, x -> r
```

In this example the Mapping was evaluated at a single point; in general Mapping's are optimized to be evaluated at an array of points.

5 Grids

Grids define a discrete representation of a mapping. The main grid classes are

1. `MappedGrid`: a grid for a single mapping that contains, among only things, a mapping and a mask array for cut-out regions.
2. `GridCollection`: A collection of MappedGrid's.
3. `CompositeGrid`: A valid overlapping grid containing interpolation information.

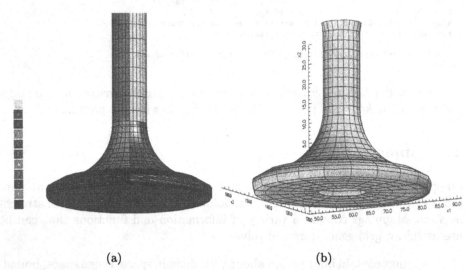

(a)   (b)

Fig. 2. Two examples of mappings. (a) A composite surface consisting of 10 trimmed NURBS (CompositeSurfaceMapping). (b) Single surface grid generated using hyperbolic grid generation (HyperbolicSurfaceMapping).

Grids contain many *geometry arrays* such as grid points, Jacobian, normals, face areas, and cell volumes. These are optionally computed as needed by solvers. In this next example we create a grid from a mapping and output the vertices.

```
AnnulusMapping annulus;          // create a mapping for an annulus
MappedGrid mg(annulus);          // Create a grid for an annulus
mg.update();                     // compute some standard geometry arrays
mg.vertex.display("Vertices (x,y)");   // The vertices (mg.vertex is an A++ array)
```

The `GridCollection` class is used to represent an overlapping grid (through the `CompositeGrid` derived class) and to represent an adaptively refined grid, figure (3).

6   Grid Functions

Grid functions represent solution values (such as velocity, density, pressure) at each point on a grid or grid-collection. There is a grid function (of float's int's or double's) corresponding to each type of grid:

1. `MappedGridFunction`: A grid function that lives on a `MappedGrid` (derived from an A++ array)
2. `GridCollectionFunction`: lives on a collection of grids.

Grid-functions are defined with up to three coordinate indices (i.e. up to 3 space dimensions) and up to 5 component indices (i.e. they can be scalars, vectors, matrices, 3-tensors,...). In this example we make a grid function and set it equal to $\cos(x)\sin(y)$ on all points of the grid.

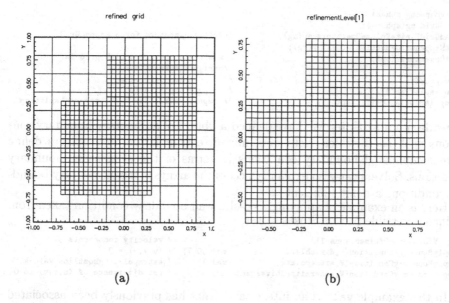

refined grid   refinementLevel[1]

(a)   (b)

Fig. 3. A `GridCollection` is used to represent an adaptive grid and its refinement levels. (a) The `GridCollection` object gc is the entire adaptive grid (3 grids). (b) The first refinement level, `gc.refinementLevel[1]` is itself a `GridCollection` (2 grids)

```
SphereMapping sphere;                          // mapping for a spherical shell
MappedGrid mg(sphere);                         // grid for a  spherical shell
mg.update();                                   // build standard grid arrays
Range all;
floatMappedGridFunction u(mg,all,all,all,2);   // define a grid function with 2 components
Index I1,I2,I3;                                // define indices for all grid points
getIndex(mg.dimension,I1,I2,I3);
// set the first component to sin(x)*cos(y)
u(I1,I2,I3,0)=sin(mg.vertex(I1,I2,I3,0))       // set the first component to sin(x)*cos(y)
              *cos(mg.vertex(I1,I2,I3,1));
```

Notice that when we declare the `floatMappedGridFunction` we do not need to indicate the the number of grid-points as this information is contained in the `MappedGrid`. The null Range `all` serves as a place holder to indicate where the coordinate directions should sit.

7 Operators

Operators define discrete approximations to differential operators and boundary conditions for grid functions. Many different types of approximations can be used. For example the class `MappedGridOperators` defines finite-difference style operators while the class `FiniteVolumeMappedGridOperators` defines cell-centred finite-volume style operators. In the following example various approximations to differential operators are computed:

```
SphereMapping sphere;
MappedGrid mg(sphere);
MappedGridFiniteVolumeOperators op(mg);        //define operators for a MappedGrid
floatMappedGridFunction u(mg), v(mg);
u.setOperators (op);                           //associate operators with grid fn.
u = ...                                        //assign u some values
v = u.grad();                                  //compute gradient of u
v = u.laplacian();                             //compute Laplacian(u)
v = op.laplacianCoefficients();                //compute matrix for the discrete Laplacian
```

Boundary conditions: We have defined a library of elementary boundary conditions such as Dirichlet, Neumann, mixed, extrapolation, etc. Solvers define more complicated boundary conditions in terms of the elementary boundary conditions. Solvers can also explicitly access the arrays if the elementary boundary conditions are not sufficient.

Here is an example of applying a no-slip wall boundary condition taken from an incompressible Navier-Stokes solver.

```
Range V(0,numberOfDimensions-1);                     // velocity components
u.applyBoundaryCondition(V,dirichlet,        wall,0.); // (u,v,w) = 0.
u.applyBoundaryCondition(V,extrapolate,      wall);    // extrapolate ghostline values
u.applyBoundaryCondition(V,generalizedDivergence,wall); // set divergence of (u,v,w) to 0
```

In this example `wall` is an integer code that has previously been associated with all the boundaries that represent no-slip walls.

8 Adaptive Mesh Refinement

AMR++ provides support for block-structured adaptive mesh refinement. It is composed of four separate libraries

LevelTransferLib : This library defines interpolation and projection operators that transfer information between refinement levels.

ParentChildSiblingLib : This library adds support for keeping track of the parent, child and sibling information associated with an adaptive grid.

RegridLib : This library will determine where to locate refinement grids in order to reduce some measure of the error in the computed solution.

SolverLib : This library provides one mechanism whereby a single grid PDE solver can be easily turned into an adaptive solver. The adaptive solver is derived from a templated base class where the single grid solver is a template parameter.

In addition to using derivation to create an AMR solver, the first 3 libraries can be directly used within an **Overture** application to add refinement capabilities. The libraries were designed so that a user can customize the algorithms if desired.

9 An Overture program to solve a simple PDE

In this example we show a basically complete code for solving the convection diffusion equation

$$u_t + au_x + bu_y = \nu(u_{xx} + u_{yy})$$

on an overlapping grid:

```
int
main()
{
  CompositeGrid cg;                    // create a composite grid
  getFromADataBase(cg,"myGrid.hdf");   // read from a data base file
  floatCompositeGridFunction u(cg);    // create a grid function
  u=1.;                                // and assign initial conditions
  CompositeGridOperators op(cg);       // create operators
  u.setOperators(cg);

  float t=0, dt=.005, a=1., b=1., nu=.1;
  for( int step=0; step<100; step++ )
  {
    u+=dt*( -a*u.x()-b*u.y()+nu*(u.xx()+u.yy()) );  // Euler time step
    t+=dt;
    u.interpolate();               // interpolate overlapping grid boundaries
    u.applyBoundaryCondition(0,dirichlet,allBoundaries,0.); // u=0 BC
    u.finishBoundaryConditions();
  }
  return 0;
}
```

10 Software availability and documentation

The **Overture** software is publicly available and can be obtained from

http://www.c3.lanl.gov/~henshaw/Overture/Overture.html.

Reports, talks and documentation are also available at this site. The introductory primer, [3], is a good place to start to learn more about the capabilities of **Overture** .

Acknowledgments: Thanks are due to the various people who have contributed to **Overture** including Kristi Brislawn, Geoffrey Chesshire, Jeffrey Hittinger, Robert Lowrie, Krister Ålander, Karen Pao, Jeffrey Saltzman and Eugene Sy.

References

1. D. L. BROWN, G. S. CHESSHIRE, W. D. HENSHAW, AND D. J. QUINLAN, *Overture: An object oriented software system for solving partial differential equations in serial and parallel environments*, in Proceedings of the Eight SIAM Conference on Parallel Processing for Scientific Computing, 1997.
2. W. HENSHAW, *Overture: An object-oriented framework for solving PDEs in moving geometries on overlapping grids using C++*, in Proceedings of the Third Symposium on Overset Composite Grid and Solution Technology, 1996.
3. ———, *A primer for writing PDE codes with Overture*, Research Report LA-UR-96-3894, Los Alamos National Laboratory, 1996.
4. D. QUINLAN, *A++/P++ class libraries*, Research Report LA-UR-95-3273, Los Alamos National Laboratory, 1995.

Optimization of Data-Parallel Field Expressions in the POOMA Framework

William Humphrey, Steve Karmesin, Federico Bassetti, and John Reynders

Los Alamos National Laboratory

Abstract. The POOMA framework is a C++ class library for the development of large-scale parallel scientific applications. POOMA's Field class implements a templated, multidimensional, data-parallel array that partitions data in a simulation domain into sub-blocks. These subdomain blocks are used on a parallel computer in data-parallel Field expressions. In this paper we describe the design of Fields, their implementation in the POOMA framework, and their performance on a Silicon Graphics Inc. Origin 2000. We focus on the aspects of the Field implementation which relate to efficient memory use and improvement of run-time performance: reducing the number of temporaries through expression templates, reducing the total memory used by compressing constant regions, and performing calculations on sparsely populated Fields by using sparse index lists.

1 Introduction

POOMA [2, 4], Parallel Object-Oriented Methods and Applications, is an object-oriented framework implemented as a set of templated C++ class libraries for scientific computing applications which require the memory and computational speed afforded only by existing and future high-performance parallel computers. An object-oriented approach is used to manage the complexity of explicit parallel programming by encapsulating data distribution and communication among processors under a predominately data-parallel interface. In this paper we describe properties of the POOMA Field class, which implements a multidimensional data-parallel array. It enables the user to write statements similar to Fortran 90 array statements. For example, this is a five-point stencil:

```
B[I][J] += A[I][J] + A[I+1][J-1] - A[I-1][J-1] - A[I+1][J+1];
```

where A and B are Fields and I and J are Index objects which represent the stencil evaluation on A accumulated into B with the tensor product domain $I \otimes J$.

POOMA evolved from, and is driven by, the needs of a diverse set of applications centered at Los Alamos National Laboratory. Simulation techniques required of POOMA include structured-mesh finite-volume methods, spectral solvers, and particle-in-cell (PIC) calculations. One particular Los Alamos application is a multimaterial hydrodynamics simulation which requires solution of a coupled system of fluid and material-property equations for compressible flow of multiple material species. The fluid and material equations are discretized on a mesh and evolved over time to perform virtual experiments. The goal is to provide detailed and accurate simulations of complex, three-dimensional systems on a variety of time scales.

A simulation in this problem domain can contain ten or more materials with widely varying properties. Both solids and gasses may be present, with realistic equations of state and elasticity/plasticity. Some materials cover much of the domain, while others will be present in only a small fraction of the total space. Due to the requirements of this application for high performance on very large problem sizes, POOMA uses several approaches to reduce the memory required to represent the material Fields, to efficiently implement the Field expressions, and to reduce the time to perform operations involving sparsely populated Fields. These are *expression templates, sparse indices,* and *compression.*

Expression Templates. Expression templates [3] are a technique for building a representation of the parse tree for the right-hand side of an expression that avoids the high cost of binary arithmetic operators. We have developed PETE, the Portable Expression Template Engine, for manipulating Fields and other objects with expression templates.

Sparse Indices. Sparse Field operations, which involve only a small fraction of the problem domain, may be carried out with greater efficiency by computing and storing the specific indices of the relevant Field elements which require manipulation, and performing operations on only this computed index set. These sparse index lists can be used as an alternative to traditional where blocks; a sparse index list is used to store the indices at which the where statement evaluates true, and is used in subsequent operations which would have been included in the body of the where block. At the cost of the extra memory needed to store the sparse index lists, the time for evaluating an expression can be reduced by the fraction of the domain included in the sparse index list.

Compression. If large subsets of a Field have some constant value, they can be represented with a single number for dramatic savings in memory and cache use. In many situations POOMA Fields can compress those regions automatically and transparently. Compression complements sparse index use by not allocating storage in areas where it is not needed, but with a coarser granularity.

The following sections describe each of these techniques in more detail, including descriptions of notation, implementation, and performance.

2 Expression Templates

Much of the mechanics of expression templates has been described in detail elsewhere ([?] includes performance comparisons); here we give only an overview.

C++ operator overloading works well for fairly small objects like complex numbers that can be stored on the stack or in registers. But for large objects like arrays allocated on the heap, the compiler cannot optimize away the intermediate temporaries created. Even a good implementation of memory management using binary operators will require several memory references for each arithmetic operation. This is an unacceptable level of inefficiency, as memory access is far slower than arithmetic operations.

Expression templates eliminate that memory overhead for an expression by not evaluating the operators immediately. Instead, an object is built that represents the entire expression, which is evaluated all at once in a single loop, with no heap-based temporaries.

Most implementations of expression templates are tightly coupled to a particular array class and impose a particular way of iterating through their elements. We have built the Portable Expression Template Engine (PETE), an efficient and generic implemention of expression templates which is retargetable to different kinds of arrays or other containers.

When PETE-aware objects appear in expressions, such as the POOMA Field expression in Fig. 1(a), the PETE system builds an object that represents the parse tree (Fig. 1(b)). In conventional expression analysis, that parse tree is typically represented by a data structure with several nodes allocated on the heap, and with pointers to each other. An expression template parse tree instead uses the C++ template mechanism to build a unique class for each expression. The entire tree is then encapsulated within a single class, where it is possible to write an evaluation function that is fully inlined and very efficient.

In PETE, the parse tree object acts like a container in the STL [1] sense, where the contained objects are operators and arrays. PETE provides a for_each function for iterating over the contents of the parse tree, and a user-supplied functor is applied at each of the leaves and internal nodes of the tree. An example of a functor which performs evaluation with integer offsets is given in Fig. 1(c). User-supplied functors could perform other tasks such as dependency checking or pointer increment. In this manner, PETE objects are concerned only with representing the parse tree, whereas the user of the parse tree decides how it is to be used. Through this division of labor, PETE has no direct dependence on the rest of POOMA, making it a separately exportable and reusable package.

POOMA evaluates each expression by looping over the appropriate domain, determining which subdomains are on the current target processor, finding which blocks are in the current expression, and then using PETE to evaluate the expression on those subdomains. For many expressions, writing POOMA Field operations using expression templates produces executables that run nearly as fast as the hand-coded C or FORTRAN equivalent. This provides the simplicity of working with a data-parallel notation in a language with all the object-oriented and template tools for building and maintaining large and complex systems.

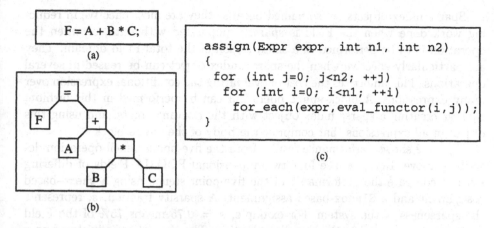

$$F = A + B * C;$$

(a)

(b)

```
assign(Expr expr, int n1, int n2)
{
  for (int j=0; j<n2; ++j)
    for (int i=0; i<n1; ++i)
      for_each(expr,eval_functor(i,j));
}
```

(c)

Fig. 1. Schematic representation of expression templates. The user types an expression like that in (a), which gets converted to a parse tree object (b), that is finally evaluated inside a single nested loop (c).

3 Sparse Indices

The POOMA framework provides a **where** statement which can be used to perform conditional assignments to a Field based on a predicate. This example sets a Field A to have values no larger than `maxval`:

```
A[I][J] = where(gt(A[I][J],maxval), maxval, A[I][J]);
```

If it is necessary to perform several operations using the same predicate, the total amount of work done can be reduced. First evaluate just the predicate at all the points in the domain, and create a list of points in the index space where the predicate is true (the *sparse index list*). Then perform subsequent operations over just the subset of the total domain stored in the sparse index list, reducing the total number of assignments to the number of sparse indices. Using this method, the sample **where** statement shown above can be expressed as:

```
SIndex<Dim> S = gt(A[I][J], maxval);
A[S] = maxval;
```

Once constructed, a sparse index (**SIndex**) object in POOMA can be used in expressions using a syntax analogous to the use of **Index** objects. Furthermore, **SIndex** objects can be combined and intersected with one another to track and manage complex sparse patterns throughout a simulation. Sparse index lists can be used in stencil operations, such as the five-point stencil discussed earlier, by including sparse index offsets in parentheses in this way:

```
B[S] += A[S] + A[S(1,-1)] - A[S(-1,-1)] - A[S(1,1)];
```

In this case, the expression will loop over the sparse index list S on the left hand side, and over the lists on the right hand side, adding the specified offsets to each index value.

Sparse index objects are so-named because they are most effective in reducing work done when the Field is sparsely populated with data, or when the operation is to be performed in a small region of the total Field domain. They are particularly effective when the sparse index object can be reused in several operations. This amortizes the cost of evaluating the conditional expression over many expressions. A block-where operation can be performed in this fashion, by first creating a sparse index object with the relevant points, and using this object in all expressions that comprise the body of the where block.

Figure 2 shows performance results from the five-point stencil operation described above. Here, we use four two-dimensional POOMA Fields of differing sizes to compare the performance of the five-point stencil using a where-based assignment and a SIndex-based assignment. A sparsity fraction, s, represents the sparseness of the system. For example, $s = 0.75$ means 75% of the Field is constant or unused in the stencil evaluation. The simulation is initiated as a constant Field with a spike of data at a single point in the center of the domain. In this way, the Field grows from a sparsity of almost 1.0 to a sparsity of 0.0 as driven by the diffusion pattern evolved by the stencil. The horizontal axis in Figure 2 gives the sparsity of the system and the vertical axis gives the relative performance of the sparse index assignment over the where assignment. The sparse index was reused 10 times (that is, the where statement was executed 10 times, and for the sparse index case, the time includes both creating the sparse index list and using it in an assignment 10 times).

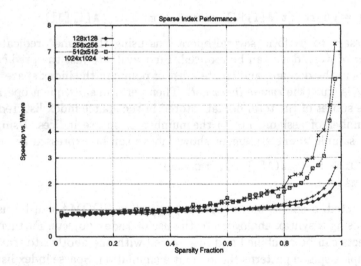

Fig. 2. Relative performance of sparse index assignment vs. the equivalent where assignment for 2-D Fields of different sizes.

Since the system is originally in a state where s is almost 1.0, the simulation progresses from right to left along the curves in Figure 2. Initially, almost all of the data is concentrated in a small region, and the use of sparse index lists noticeably reduces the total time needed to compute the stencils, by up to a factor of three or more when $s \approx 0.9$. As would be expected, when the system evolves to cover most of the domain, the effectiveness of using sparse index lists decreases; in this example, the point where no improvement is seen anymore occurs near $s \approx 0.4$. The point where this occurs will vary with problem size and with the number of times the sparse index list can be reused.

4 Compression

In some simulations, there will be small active regions of interest and large inactive regions which remain constant. The location of a material or physics event which only exists in a small part of the simulation domain is an example of such a situation. Data-parallel constructs, such as arrays or fields, are not well-suited to such cases, since memory and CPU time are spent in uninteresting parts of the simulation. If the regions are static, it may be possible to allocate Field subdomains in a patchwork to tile a region of interest. However, in cases where the location of the active region is dynamic such as material interface tracking or wave propogation, another approach is required.

Deciding what part of the domain is "uninteresting" can be very complex, and POOMA Fields cannot determine this a priori. However, POOMA can detect parts of the domain where the value of a Field is constant, and operate more efficiently in those regions by compressing the storage. A Field with some regions compressed then acts in every way like an uncompressed Field, except that it is more efficient in both memory usage and operations performed.

4.1 Expressions with Compression

In the current implementation of POOMA, each Field is broken into subdomains referred to as virtual nodes or "vnodes". Each vnode represents a rectangular subset of the whole domain, with constant strides in each dimension. A vnode can be compressed when all of its elements have the same value. In that case, storage for only one value needs to be allocated, with the strides set to zero. A vnode can be compressed by examining every element but, as we will see below, this can sometimes be done automatically. The number of virtual nodes assigned to a processor in a parallel simulation can vary depending upon the granularity of compression desired.

Consider the expression

```
B[I][J] += A[I][J];
```

where the pattern of vnode domains is the same for each of A and B, but the compression patterns are different, as shown in Fig. 3(a) and Fig. 3(b). The compression pattern for B after this assignment can be determined entirely from the compression patterns of B and A before the assignment, and is shown in Fig. 3(c). For this example, there are four cases for the combinations of vnodes,

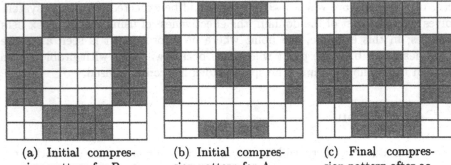

(a) Initial compression pattern for B.

(b) Initial compression pattern for A.

(c) Final compression pattern after accumulating A into B.

Fig. 3. Update of the compression pattern in a Field for an accumulation assignment. The shaded blocks are uncompressed, and the unshaded are compressed. After accumulation at most the union of the patterns are uncompressed.

depending on whether the left-hand and right-hand sides are compressed or uncompressed, summarized in Table 1. In this table, 'c' means compressed, and 'u' means uncompressed. 'N' evaluations means that the right-hand side is evaluated for all the points in the vnode, and '1' means it only needs to be evaluated once. '2N' memory references means that the accumulation must read from each element of the right and write to each on the left, 'N+1' means it must write to each on the left but read from only one element on the right, and '2' means it reads once from the right and writes once on the left.

Table 1. Vnode compression options.

Initial Arrays		operator +=			operator =		
B	A	Final B	Evals	Memory Refs	Final B	Evals	Memory Refs
c	c	c	1	2	c	1	2
c	u	u	N	2N	u	N	2N
u	c	u	N	N+1	c	1	2
u	u	u	N	2N	u	N	2N

If the operation is a simple assignment rather than an accumulation, the logic is slightly different, since the contents of B would not be read, only written. Then the pattern of compression in B after the assignment would be the same as in A. The operations performed in this case are also shown in Table 1.

In more complex expressions, cancellation of terms could lead to a case where an uncompressed vnode has constant data throughout. In these cases, a Field must be explicitly instructed by the user to look for and compress vnodes with constant values. It would be inefficient to scan the contents of Fields automatically to see if they have become constant after each assignment. Thus, in actual simulations, Fields are periodically checked for possible compression. Compression provides a handy technique for reclaiming memory from global temporary Fields. By assigning a constant to a Field, the entire Field compresses and acts as a lightweight proxy until uncompressable data is actually written into it.

4.2 Compression Performance

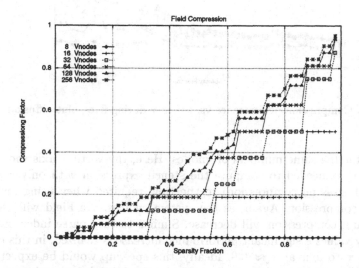

Fig. 4. Compression factor vs. sparsity during five-point diffusion.

Using the five-point diffusion example described earlier, the evolution of the compression factor—the fraction of elements in compressed vnodes relative to the total number of elements in the system—was computed over the course of a simulation similar to that used for the sparse index benchmarks. The simulation begins with all Field elements constant, except for a single spike in the center of the domain, so initially the system has very large sparsity (s almost 1.0). This is the maximum compression that it can have, since the system diffuses outward from the center, uncompressing vnodes as they are encountered. Figure 4 shows the sparsity vs. compression factor for six different numbers of vnodes on a 512x512 Field for this 2-D diffusion example. For small enough vnodes, this would be a straight line with a slope of one, which would be the optimal case. For coarse vnode blocking, however, we see discontinuous drops in the compression factor as the system spreads out, due to the sudden expansions which occur when two or more elements in a vnode take on different values. As the number of vnodes increases, the curves in Figure 4 approach the asymptotic case.

Figure 5 shows the improvement in run-time performance which results from the use of compression for the same 512x512 2-D diffusion example shown in

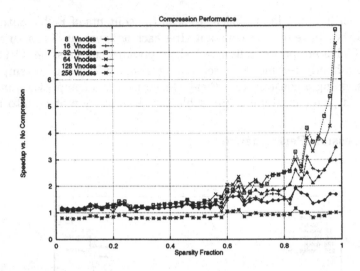

Fig. 5. Compression performance vs. sparsity during five-point diffusion

Fig. 4, and for the same numbers of vnodes. Here, the vertical axis shows the ratio of the time needed to compute the stencil expression with only a single vnode (which uses no compression) to the time required when using multiple vnodes and compression. Again, as the system evolves, the Field will become less sparse and compression will decrease. Similar to the sparse index case, at high sparsity values a significant performance increase is realized, in this case a factor of three to four at $s \approx 0.9$. Ideally, this speedup would be expected to be on the order of $1/(1 - s) = 10$, but this is not realized due to the overhead of looping through the vnodes, and the finite vnode sizes as shown in Figure 4. The optimal improvement in the system shown here occurs with 32 or 64 vnodes on a 512x512 Field; for larger numbers of vnodes, the performance actually decreases. For any problem size and memory architecture, there is an optimal vnode granularity for compression that must be determined.

5 Conclusions

Data-parallel operations are an effective approach to representing physics abstractions at a high level, and encapsulating the details of parallel simulation. Field operations suffer performance degradation in expressions based on overloaded operators, and also in cases where data is localized to a small region of the simulation domain or is sparsely distributed. Three techniques to address these issues were presented: the Portable Expression Template Engine (PETE), sparse indexing, and Field compression. These techniques were described in the context of implementation within the POOMA framework. It was shown that these techniques could each be used to help improve the performance of applications with Field expressions across compressed regions and sparse distributions, if care is taken in choosing the proper domain of applicability when applying the methods.

Acknowledgements This research was performed at the Advanced Computing Laboratory of Los Alamos National Laboratory, Los Alamos, NM under the auspices of the United States Department of Energy. The authors would especially like to thank the Department of Energy for support through the DOE 2000 Scientific Template Library project, and all of the members of the POOMA team for their help in this work.

References

1. David R. Musser and Atul Saini. *STL Tutorial and Reference Guide: C++ Programming with the Standard Template Library.* Addison-Wesley, 1996.
2. John Reynders et al. Pooma: A framework for scientific simulations on parallel architectures. In Gregory V. Wilson and Paul Lu, editors, *Parallel Programming using C++*, pages 553–594. MIT Press, 1996.
3. Todd Veldhuizen. Expression templates. Technical Report 5, C++ Report 7, June 1995.
4. Tim Williams, John Reynders, and William Humphrey. POOMA User Guide. Advanced Computing Laboratory - Los Alamos National Laboratory, April 1997. http://www.acl.lanl.gov/pooma/doc/userguide/.

MC++ and a Transport Physics Framework

Stephen R. Lee[1], Julian C. Cummings[1], Steven D. Nolen[2], and Noel D. Keen[3]

[1] Los Alamos National Laboratory
[2] Texas A&M University
[3] University of New Mexico

1 Motivation

The U.S. Department of Energy has launched the Accelerated Strategic Computing Initiative (ASCI) to address a pressing need for more comprehensive computer simulation capabilities in the area of nuclear weapons safety and reliability. Since the U. S. government has decided to abandon underground nuclear testing, the Science-Based Stockpile Stewardship program is focused on using computer modeling to assure the continued safety and effectiveness of the nuclear stockpile. Doing this will require major advances in computing speed and in the complexity of computational models. Thus it is anticipated that more modern software development approaches will be needed to accelerate prototyping of new computer applications, facilitate the integration of different physics models, manage the complexity of simulation codes, and port the simulation codes to new computational platforms.

We believe that the use of object-oriented design and programming techniques can help in this regard. Object-oriented programming (OOP) has become a popular model in the general software community for several reasons. It can help manage a software application by encouraging code designers to break the application down into constituent *objects* and their interactions. This leads to a natural decomposition of a software project into manageable pieces. Often these pieces may have multiple uses within a given project or may be reused in subsequent projects, thus shortening the time for application development. Finally, object-oriented programming provides for explicit interfaces between relatively self-contained objects, which typically makes adding new code or extending an existing application more straightforward.

These benefits transfer very well to the realm of high-performance physics modeling and rapid application development required for ASCI. Many people will be lending their expertise to ASCI application development, and OOP will help them work together more effectively. The simulation codes produced must be flexible and agile, quickly adapted to the newest high-performance computing platforms as they become available. Moreover, these codes must be written in a high-level, easily understood fashion, so that the critical physics knowledge embedded in the models is not lost as codes are passed along from one researcher to the next. OOP can help address these important aspects of ASCI code development.

MC++ is an ASCI-relevant application project that demonstrates the effectiveness of the object-oriented approach. It is a Monte Carlo neutron transport code written in C++. It is designed to be simple yet flexible, with the ability to quickly introduce new numerical algorithms or representations of the physics into the code. MC++ is easily ported to various Unix workstations and parallel computers, such as the three new ASCI systems, largely because it makes extensive use of classes from the Parallel Object-Oriented Methods and Applications (POOMA) C++ class library. The MC++ code has been successfully benchmarked using some simple physics test problems, has been shown to provide comparable performance to that of a well-known Monte Carlo neutronics package written in Fortran, and was the first ASCI-relevant application to run in parallel on all three ASCI computers.

2   Algorithm

MC++ computes the k and α eigenvalues of the neutron transport equation. These eigenvalues are used to assess criticality of a system containing fissile material. Such a system is said to be *critical* if there is a self-sustaining, time-independent chain reaction in the absence of an external source of neutrons. In a critical system, the rate of neutron production via fission is just equal to the rate of loss due to absorption and leakage from the system. If there is no such equilibrium, then the neutron population will either increase (a *supercritical* system) or decrease (a *subcritical* system) exponentially in time. Each of the eigenvalues treats this criticality problem in a slightly different way, but both provide valuable information on the criticality of the system in question.

MC++ performs its calculations on a three-dimensional Cartesian mesh. Information on the dimensions of the mesh, the types of materials contained in the system, and material densities in each mesh cell are obtained from another simulation code and read in to MC++. It stores this information along with a database of isotopic information supplied by the user to describe the materials in the problem. Then, the Monte Carlo calculation is begun by loading neutrons into cells containing fissile material in a "round robin" fashion. The neutrons are tracked through the system, undergoing collisions with isotopes that compose the materials in each mesh cell and boundary interactions with the mesh itself. Collisions that result in a fission event produce new neutrons which are stored as source points for the next generation or cycle. All particles are tracked until they trigger a fission event, are absorbed, or escape from the problem. Once all particles in the current generation have completed their tracks, the next generation is begun. During each cycle, MC++ accumulates tallies of specific events and uses this information to compute estimates of k or α.

3   Implementation

The MC++ code is largely written in an object-oriented, data-parallel style, and it makes extensive use of classes from the POOMA Framework[1, 2]. POOMA is a C++ class library designed to encapsulate the details of parallel programming and provide a set of high-level, physics-based data structures and algorithms with which to build scientific applications. In object-oriented programming, a general structure or outline for a set of codes intended to address a certain class of problems is known as a *framework*. The aim of the POOMA class library is to be an object-oriented framework for scientific computing on parallel architectures. Using this framework, we have built a simple and efficient Monte Carlo neutronics package.

The most heavily used POOMA feature in MC++ is its particle simulation capabilities. POOMA provides a `ParticleBase` class with a minimal description of a particle population containing only a position and an identification number for each particle. The user (in this case, the MC++ application developer) then derives a class from `ParticleBase` and adds data members to describe the other particle characteristics, such as velocity or mass, needed for this simulation model. This method allows for an extremely flexible particle description. Nevertheless, the `ParticleBase` class provides the derived class with many powerful features, including automatic decomposition of the particle data across processors, the ability to add or delete particles from the particle population at will, and a convenient *array syntax* for performing data-parallel operations involving particle attributes.

MC++ also makes use of the `Field` and `Mesh` types in POOMA. It stores the mesh description inside a POOMA `Cartesian` object, which represents a non-uniform Cartesian mesh. The `Cartesian` class provides a translation from "index space" to the physical domain of the simulation, and simplifies the calculation of the distance to the nearest mesh cell boundary performed in MC++. POOMA `Field` objects are used to hold information on the type and density of each material in each mesh cell. The data is automatically distributed across processors, so that only a portion of the data of each `Field` is owned by each processor. This allows MC++ to perform simulations on very large meshes with a relatively small number of processors. POOMA has a data layout option which will maintain data for each particle on the same processor that owns data for that portion of the field near the particle's current location. Maintaining this *data locality* allows for much more efficient computations and lookups of field data. Although MC++ mainly uses the POOMA `Field` as a storage class, it has many other powerful features, such as built-in boundary conditions, array syntax for writing expressions involving `Fields`, and optimized differential operators like `Div()` and `Grad()` for `Fields` residing on a specific type of POOMA `Mesh`.

Another major feature of the latest version of POOMA is the load balancer (see Fig. 1). In MC++, a Particles object is passed to the `BinaryRepartition` function. This function extracts the particle position attribute, creates an empty `Field` on the current `FieldLayout`, and does a nearest-grid-point scatter of the particle number density onto the `Field`. Then it calls a function to calculate a new set of subdomains which will distribute the values of this "weight" `Field` as evenly as possible. This is done recursively, starting with the entire domain.

Load Balancing

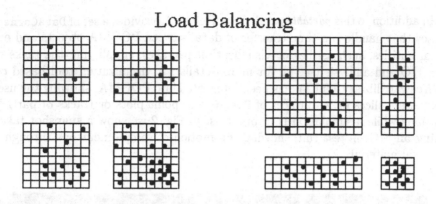

Fig. 1. Schematic of load balancing procedure in POOMA.

The axis containing greatest number of cells is located and labeled as the cut axis. The weight Fields along each axis (except the cut axis) are summed, yielding a one-dimensional array of reduced weights with the same length as the cut axis. The median of this 1D array, where the running sum of the reduced weights is closest to half of the total sum, is the cut location, and the domain is split here, yielding two subdomains. This process is then repeated recursively on the subdomains $\log_2 N$ times to generate N subdomains. The entire process is parallelized as much as possible by distributing the subdomains to be cut in each recursion among processors. When the new set of subdomains has been obtained, it is used to repartition the original FieldLayout. The FieldLayout then tells each of its "users" (i.e., Fields and ParticleLayouts) to redistribute their data.

Other features of POOMA that add value to the MC++ application are worthy of mention. POOMA manages virtually all of the parallelism and other "computer science" issues that arise in getting this neutronics code to run efficiently on various architectures. The only explicit message passing in MC++ is done in the Tally classes to gather event tallies across all the processors. (The Tally classes were built specifically for Monte Carlo simulation and were not provided by POOMA.) Everything else, including the maintenance of particle data locality and coordination of operations on the distributed particle data, is done transparently by POOMA. This allows the vast majority of MC++ to be written using high-level objects in an easily understood syntax that clarifies the physics content while hiding the computer science details. Furthermore, the extensive use of C++ templates and the *expression template* technique[3] in POOMA helps the compiler to produce highly optimized code with little or no sacrifice in application performance, while providing a high-level application development syntax.

In addition to this *portable parallelism*, POOMA provides a set of `DataConnect` classes that handle run-time transfer of data between POOMA objects and external entities, such as files or visualization packages, simplifying the tasks of data I/O and analysis (see [4] for more details). A visualization tool based on the freeware library VTK[5] has been integrated into POOMA, allowing the user to generate slices or isosurfaces of `Fields` and point plots or tracks of particle positions while a simulation is in progress. In Fig. 2 we show a snapshot taken during an MC++ test run, showing the motion of sample neutrons through a spherical material.

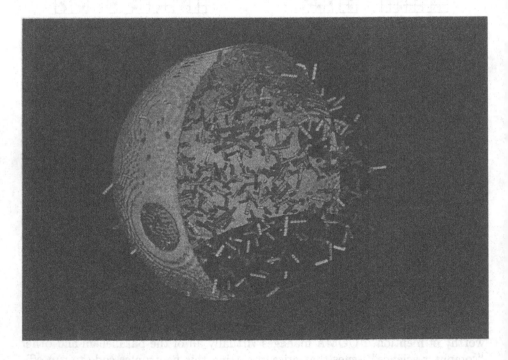

Fig. 2. Neutron tracks (colored by kinetic energy) through a bare uranium sphere.

4 Code Results

The transport eigenvalue estimates produced by MC++ have been benchmarked for accuracy against the MCNP code for a set of simple test problems. MCNP is written in Fortran 77 with PVM for message passing on parallel computers, and uses an analytic description of materials in the simulated system rather than a mesh description. MC++ provides eigenvalues within statistical error of those provided by MCNP for the standard transport benchmark problem "Godiva", which is a bare uranium sphere.

The code performance of MC++ has also been compared with MCNP on a variety of computing platforms, including several types of Unix workstations and the Cray T3D. To make the comparison with MCNP fair, mesh cell boundary crossing events were turned off in MC++, effectively making the selected problem an infinite-medium problem (which implies no material or mesh boundary crossings). For this early comparison, MC++ showed effectively identical serial performance to MCNP on all tested platforms, and vastly superior parallel performance on the T3D, the only modern parallel computing platform on which MCNP was readily available. MC++ has been rapidly ported to many different types of parallel computers, including the Intel Teraflops machine (ASCI Red), an SGI Origin 2000 cluster (ASCI Blue Mountain), and the IBM SP2 (ASCI Blue Pacific). Table 1 shows all platforms on which MC++ has been run. Platforms shown in bold have parallel capability.

Table 1. MC++ platforms.

Platform	Description
TFLOP	Intel TeraFlop (ASCI Red)
ORIGIN	SGI Origin 2000 (ASCI Blue Mountain)
SGIO2	SGI O2 Workstation
SP2	IBM SP2 (ASCI Blue Pacific)
T3D	Cray T3D
RS6K	IBM RS6000 Workstation Cluster
SUN4SOL2	Sun Workstation w/Solaris OS

MC++ showed reasonably good parallel efficiency (within 70–80% of linear speedup) up to 8 nodes on all platforms.[1] No special performance tuning of MC++ was required on any platform. See Fig. 3 for representative parallel speedups on the ASCI platforms.

5 Transport Physics Framework

An area of continuing development in the MC++ project is the generalization of the MC++ code into a Transport Physics Framework (TPF). Just as the POOMA framework has been quite useful in accelerating code development and promoting code reuse across applications, a TPF could aid in the development of new transport physics codes and facilitate testing of novel techniques. What is needed are a general outline of the process of transport simulation, and a set

---

[1] With more than 8 nodes, load balancing becomes important in this 3D test problem. Parallel efficiency gains in MC++ due to the use of load balancing are currently being explored.

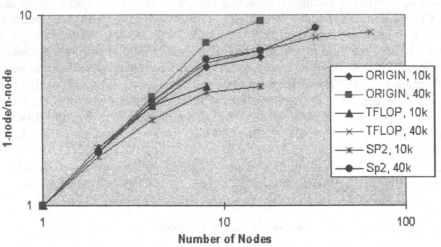

Fig. 3. Parallel performance of MC++ on ASCI platforms for Godiva problem, on a 128x128x128 mesh, with 10000 or 40000 particles.

of tools that could be directly used or readily modified for a specific simulation need.

MC++ contains a variety of TPF-relevant abstractions. For example, MC++ uses a set of cross-section classes to encapsulate the type of cross-section data used during transport and to make the reading of different cross-section data libraries, the storage of the data, and the retrieval of required data during transport simple and general. MC++ was initially developed to have full multi-group capabilities. However, continuous energy cross sections, which are completely different in scope and nature from multi-group cross sections, were recently added to MC++, along with the appropriate physics capabilities for using the new data. Through inheritance of a new cross-section class from the existing abstractions, the addition of this new capability was straightforward.

MC++ also uses a set of particle sourcing classes representing various physics models for direction, energy, and angular distribution. This set of abstractions allows one to trivially add new particle sourcing methods to the code, and have them be automatically used serially or in parallel, with the transport algorithms remaining unchanged.

It is invaluable to have such methods readily available in the form of interchangable objects with a common interface. Such a "plug and play" capability will greatly simplify the task of evaluating the effectiveness and utility of the various algorithms. This same sort of generality would be beneficial in other areas of MC++, such as the modeling of neutron capture events and the representation of material isotopic information. Reformulating MC++ in terms of a TPF will provide the ASCI project and other researchers with a much more flexible tool for the exploration of alternative techniques and models in neutron transport studies.

References

1. Tim Williams, John Reynders, and William Humphrey. *POOMA User Guide*. Advanced Computing Laboratory — Los Alamos National Laboratory, 1997. http://www.acl.lanl.gov/pooma/doc/userguide.
2. J.V.W. Reynders et al. Pooma: A framework for scientific simulation on parallel architectures. In Gregory V. Wilson and Paul Lu, editors, *Parallel Programming using C++*, chapter 16, pages 553–594. MIT Press, 1996.
3. Todd Veldhuizen. Expression templates. *C++ Report*, 7(5), June 1995.
4. William Humphrey and James Ahrens. *POOMA External Data Connection Guide*. Advanced Computing Laboratory — Los Alamos National Laboratory, 1997. http://www.acl.lanl.gov/pooma/doc/externguide.
5. Will Schroeder, Ken Martin, and Bill Lorensen. *The Visualization Toolkit: An Object-Oriented Approach to 3D Graphics*. Prentice Hall, 1996.

The Role of Abstraction in High-Performance Computing*

Brian C. McCandless and Andrew Lumsdaine

University of Notre Dame

Abstract. Although there is perpetual interest in using high-level languages such as C++ for high-performance computing, the conventional wisdom is that the very data abstractions that make these languages attractive from a software engineering perspective carry with them inherent performance penalties that make them unattractive from a performance perspective. The Matrix Template Library (MTL) is a C++ library specification that consists of a small number of composable template classes for defining sparse and dense matrix types as well as a comprehensive set of generic algorithms for numerical linear algebra. In this paper, we discuss our experiences with MTL and demonstrate that abstraction is not necessarily the enemy of performance and that, in fact, data abstraction can be an effective tool in enabling high performance.

1 Introduction

There is a common perception in scientific computing that abstraction is the enemy of performance. Although there is perpetual interest in using languages such as C or C++ and the powerful data abstractions that those languages provide, the conventional wisdom is that data abstractions inherently carry with them a (perhaps severe) performance penalty. Our thesis is that this is not necessarily the case and that, in fact, abstraction can be an effective tool in enabling high performance—but one must choose the right abstractions.

The misperception about abstraction springs from numerous examples of C++ libraries that provide a nice user interface through polymorphism, operator overloading and so forth, so that the user can implement an algorithm or a library in a "natural" way (SparseLib++ and IML++ [4], for example). Such an approach will (by design) hide computational costs from the user and degrade performance. One approach to providing performance and abstraction is through the use of lazy evaluation [1], but this approach can have other performance penalties as well as implementation difficulties.

One of the most important concerns in obtaining high performance on modern workstations is proper exploitation of the memory hierarchy. That is, a high-performance algorithm must be cognizant of the costs of memory accesses and must be structured to maximize use of registers and cache and to minimize

* This work was supported by NSF cooperative grant ASC94-22380.

cache misses and pipeline stalls. To properly complement high-performance algorithms, data abstractions should similarly account for hierarchical memory explicitly, and enable a programmer to readily exploit it.

The particular set of abstractions for bridging the performance-abstraction gap that we present here is the *Matrix Template Library* (MTL), written in C++ [8]. In the following sections, we describe the basic design of MTL and discuss our experiences in developing and using it. Experimental results are presented that show that MTL provides performance competitive with (or better than) traditional mathematical libraries.

We remark that this work is decidedly *not* an attempt to "prove" that a particular language (in our case, C++) offers higher performance than another language (Fortran, for example). Such arguments are, ultimately, pointless. Any language with a mature compiler can offer high performance (see PhiPAC [3]). Software development, even scientific software development, is about more than just performance and, except for academic situations, one must necessarily be concerned with the costs of software over its entire life-cycle. Thus, we contend that the only relevant discussion to have about languages is how particular languages enable the robust construction, maintenance, and evolution of complex software systems. In that light, modern high-level languages have a distinct advantage: most of them were designed specifically for the development of complex software systems, and the more widely-used ones have survived only because they are able to meet the needs of software developers.

2 The Matrix Template Library

MTL is by no means the first attempt to bring abstraction to scientific programming (see [2], for example), nor is it the first attempt at a mathematical library in C++ (see HPC++ [14], LAPACK++ [5], SparseLib++/IML++ [4], and the Template Numerical Toolkit [11]). MTL is unique, however, in its general underlying approach to separate algorithms from data structures, and in its particular commitment to self-contained high performance. Other libraries, if they are concerned about performance at all, attain high performance by making (mixed-language) calls to BLAS subroutines. The higher-level C++ code merely provides a syntactically pleasing means for gluing high-performance subroutines together, but does not provide flexible means for obtaining high performance (as MTL does).

The design of MTL is partly based on the idea that linear algebra objects (matrices and vectors) can be composed from simpler components. This can lead to good reuse of code, flexible storage formats, and easy-to-manage data distributions. The design is also based on the premise that storage format and algorithms can be efficiently separated and treated independently for a large class of algorithms. As a result, each algorithm needs to be implemented once and can automatically be called with matrices and vectors in any storage format. This drastically reduces the amount of code that needs to be written, debugged, and maintained.

3 Data Structures and Algorithms

One design goal of MTL is to separate algorithms from local data storage formats and from the distribution of matrices and vectors. This separation is a key ingredient in making the library extensible in both the number of linear algebra routines and the number of supported data storage formats.

The Matrix Template Library was inspired to a large extent by the Standard Template Library (STL) for C++ [7]. STL has become extremely popular because of its elegance, richness, and versatility. The original motivation for STL, however, was not to provide yet another library of standard components, but rather to introduce a new programming paradigm [9, 10].

This new paradigm was based on the observation that many algorithms can be abstracted away from the particular representations of the data structures upon which they operate. As long as the data structures provide a standard interface for algorithms to use, algorithms and data structures can be freely mixed and matched. Moreover, this paradigm accommodates this process of abstraction without sacrificing performance.

To realize an implementation of an algorithm which is independent of data structure representation requires language support. In particular, a language must allow algorithms (and data) to be parameterized not only by the values of the formal parameters (the arguments), but also by the *type of the data*. Few languages offer this capability, and it has only (relatively) lately become part of C++. In C++, functions and object classes are parameterized through the use of *templates* [13], hence the realization of a generic algorithm library in C++ as the Standard Template Library.

The real contribution of STL, then, was to realize that algorithms and data structures can be separated, to classify types of standard components, to define a set of interfaces for standard component to use, and lastly, to provide a model implementation of generic algorithms and standard components in C++. Although it is this last (concrete) contribution that is most widely celebrated, the true value of STL is more profound. In developing MTL, we attempted to follow the complete theme of STL, rather than simply reconstructing STL in a linear algebra guise.

4 Components and Contracts

Component classes in MTL are designed to be interchangeable. Each component provides the same functionality and public interface as the other components of the same type. This interface is referred to as the component contract, and describes the basic requirements of each component. The semantics and syntax of each function is the same for each component, but is carried out in importantly different ways.

For example, matrices in MTL are described in terms of the local data storage format in conjunction with global mathematical properties. Distributed matrices are additionally described with a distribution. For example, the (sequential) matrix class, `Matrix` and the distributed matrix class, `DMatrix`, are described as:

```
template <class Data, class Desc> class Matrix;
template <class Dist, class Data, class Desc> class DMatrix;
```

Here, the `Data` class stores all the data of a sequential matrix, or the local data of a distributed matrix. The `Data` class contains functions for constructing, destroying, setting, getting, and iterating over the matrix data. Examples of classes that fulfill the `Data` component contract are: compressed row, compressed column, coordinate, dense row major, dense column major, blocked compressed row, and others. This list can be extended by any class which conforms to the `Data` component contract.

Similarly, the `Desc` (description) class is used to control and manage global matrix properties such as symmetry and structure (e.g., triangular, banded, general).

The `Dist` (distribution) class provides methods for mapping global matrix indices to processors and local matrix indices to control how the matrix is distributed. The same distribution classes can also be used to distribute vectors (which is another example of component reuse). Examples include block cyclic, block, and cyclic.

The `Matrix` and `DMatrix` classes provide a uniform interface to the user for manipulating, querying, and operating on the matrix data through its public interface. Internally, the two matrix classes make calls into their component classes to carry out the matrix operations.

A large number of different types can be represented merely by combining different components together. The total number of matrix storage formats is the product of the number of components within each component type, whereas the number that need to be implemented is simply the sum. This is in contrast to some mathematical libraries, particularly those that use inheritance, which must implement a new class for each matrix type. There are also different levels at which composition can take place—some categories of data storage types can be composed from simpler components. For example, a two-dimensional dense data class component can be combined with a column-orientation component to represent a BLAS general matrix. Furthermore, some two-dimensional data classes can be constructed from arrays of one-dimensional containers. Such constructs are used in many sparse types, such as those based on balanced trees or linked lists. The idea of composition is clearly a powerful one. It allows code to be reused in a number of different situations while also making each individual component small and easy to manage.

5 Iterators

One approach to achieving independence of algorithms and data structures is through the use of iterators (the approach used by STL). Iterators are objects that generalize access to other objects (iterators are sometimes called "generalized pointers"). The definition of the iterator classes in STL provide the uniform interface between algorithms and containers necessary to enable genericity. That is, each container class has certain iterators that can be used to access and perhaps manipulate its contents. STL algorithms are in turn written solely in terms of iterators

This scheme works well for one-dimensional containers, such as those found in STL. However, it has some serious drawbacks for two-dimensional matrix objects, since different storage formats have different performance characteristics depending on the order in which the elements are accessed. So, while it is possible to construct generic algorithms using one-dimensional container iterators, such an approach will tend to strongly bias one storage format over the other (or add additional inefficiencies).

For example, a relatively efficient way to implement the matrix-vector product routine using iterators is shown below. The increment operator function retrieves the element from the matrix which is closest in memory. In a compressed row storage format, the elements on the same row are closest to each other; the reverse is true for compressed column.

```
template <class MAT, class VEC1, class VEC2>
void matvec_product(const MAT &A, const VEC1 &x, VEC2 &y)
{
    for (MAT::iterator i = A.begin(); i != A.end(); ++i)
        y[i->row] += i->value * x[i->col];
}
```

However, this approach does not lend itself to compiler optimizations such as loop unrolling, nor does it efficiently handle dense and block sparse matrices. To obtain high performance while separating data representations from algorithmic implementation, new approaches are needed.

One mechanism for achieving this interface is as follows. Each storage format type implements a function, called `Iterate()`, which is templated on a linear algebra operation object. The `Iterate()` function traverses through the elements of the matrix in the most efficient way. For each new element it calls an inlined function on the linear algebra operation that performs the inner loop of the operation. The object associated with the matrix-vector product operation might look like Fig. 1. The `template<class OP> Iterate()` method is now implemented by each of the data storage formats. It efficiently iterates over the matrix, calling the `inner()` method on the linear algebra object OP for each element. Loop unrolling, register blocking, and other optimizations can be used to improve performance. It has been our experience that some compilers optimize this code very well with no performance penalty over specifically tuned code. Unfortunately, other compilers do not optimize this quite as well.

A second approach adopted by MTL is to provide interfaces to the data storage classes that are of appropriate granularity for high performance. The stored data is thus not necessarily traversed element by element, but in larger, multi-element, chunks. The specifics of the dimensions of the traversal can be

```
template <class VEC1, class VEC2, class PR>
class matvec_product_op {
public:
  matvec_product_op(VEC1& _x, VEC2& _y) : x(_x), y(_y) {}
  inline void inner(PR& val, int row, int col) {
    y[row] += val * x[col];
  }
protected:
  VEC1 &x;
  VEC2 &y;
};
```

Fig. 1. Object associated with matrix-vector product.

varied according to the algorithm requiring the traversal. For example, matrix-matrix multiplication benefits greatly from using multiple rows of a matrix.

The results in Fig. 2 show how this second approach compares with the element-wise approach in other linear algebra libraries (the NIST Sparse BLAS [12] and the Template Numerical Toolkit [11]). The test matrices come from standard collections and are each on the order of 10,000 rows and columns. For these test cases, MTL significantly outperforms the other libraries, while still providing a powerful and flexible set of abstractions.

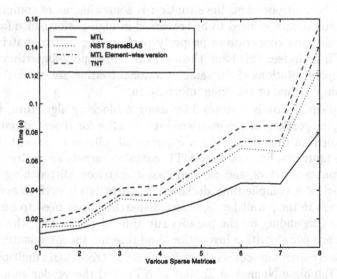

Fig. 2. Execution times for sparse matrix-vector multiplication on Sun UltraSPARC 170E.

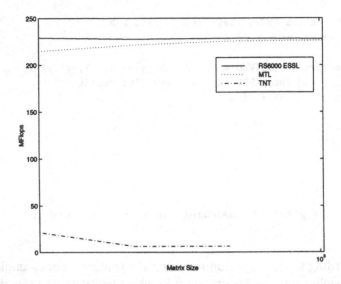

Fig. 3. MFLOPS for dense matrix-matrix multiplication on IBM RS6000.

6 Poly-Algorithms

The approaches described above for high performance are appropriate when there is a single "dominant" storage type. For example, in a matrix-vector product, the layout of the matrix dictates the structure of memory accesses. For matrix-matrix multiplication, the situation is somewhat more complicated, because two stored entities need to be traversed in high-performance fashion, and the traversals must cooperate to properly perform the matrix-matrix multiplication. MTL addresses this issue through the use of poly-algorithms, whereby different implementations of the same functional routine are invoked depending on some compile-time or run-time information.

The sequential case is addressed by using a blocking algorithm, which converts or copies portions of the matrices into a buffer for repeated reuse [6]. The blocked algorithm then dispatches a matrix multiplication kernel based on the type of the matrices. For example, MTL contains kernels for multiplying dense matrices, sparse matrices, and blocked sparse matrices. Dispatching the appropriate kernel is a compile-time decision, so there is no performance penalty. However, even in the parallel case, where dispatching may need to be performed at run time (depending on the parallel run-time environment), the dispatch is only done once per algorithm invocation, and thus has inconsequential overhead.

Figure 3 compares execution times for dense matrix-matrix multiplication for MTL, the Template Numerical Toolkit (TNT), and the vendor supplied BLAS (ESSL) on an IBM RS6000 Model 590. Although MTL is completely written in C++, it achieves nearly the performance of ESSL (recognized as being one of the best vendor libraries available).

7 Conclusions

Our initial experiences with MTL are encouraging and have borne out our conjecture that abstraction and high performance are not necessarily mutually exclusive. However, our studies with MTL thus far have been of an abstract benchmark nature. Although such studies have some value, one of the goals of MTL is to ease the development of large-scale scientific software. Future work will therefore focus on incorporating MTL into real application codes.

References

1. Susan Atlas et al. POOMA: A high performance distributed simulation environment for scientific applications. In *Proceedings Supercomputing '95*, 1995.
2. Satish Balay, William D. Gropp, Lois Curfman McInnes, and Barry F. Smith. Efficient management of parallelism in object-oriented numerical software libraries. In E. Arge, A. M. Bruaset, and H. P. Langtangen, editors, *Modern Software Tools in Scientific Computing*. Birkhauser, 1997.
3. J. Bilmes, K. Asanovic, J. Demmel, D. Lam, and C.-W. Chin. Optimizing matrix multiply using PHiPAC: A portable, high-performance, ANSI C coding methodology. Technical Report CS-96-326, University of Tennessee, May 1996. Also available as LAPACK working note 111.
4. J. Dongarra, Andrew Lumsdaine, Xinhui Niu, Roldan Pozo, and Karin Remington. A sparse matrix library in C++ for high performance architectures. In *Proceedings Object Oriented Numerics Conference*, Sun River, OR, 1994.
5. J. Dongarra, R. Pozo, and D. Walker. LAPACK++: A design overview of object-oriented extensions for high performance linear algebra. In *Proceedings of Supercomputing '93*, pages 162–171. IEEE Press, 1993.
6. Monica S. Lam, Edward E. Rothberg, and Michael E. Wolf. The cache performance and optimizations of blocked algorithms. In *ASPLOS-IV Proceedings - Fourth International Conference on Architectural Support for Programming Languages and Operating Systems*. ACM Press, 1991.
7. Meng Lee and Alexander Stepanov. The standard template library. Technical report, HP Laboratories, February 1995.
8. Andrew Lumsdaine and Brian McCandless. Parallel extensions to the matrix template library. In *Proc. 8th SIAM Conference on Parallel Processing for Scientific Computing*. SIAM, 1997.
9. David R. Musser and Alexander A. Stepanov. Generic programming. In *Lecture Notes in Computer Science 358*, pages 13–25. Springer-Verlag, 1989.
10. David R. Musser and Alexander A. Stepanov. Algorithm-oriented generic libraries. *Software-Practice and Experience*, 24(7):623–642, July 1994.
11. Roldan Pozo. Template numerical toolkit for linear algebra: high performance programming with C++ and the standard template library. In *Proceedings ETPSC III*, August 1996.
12. Karen A. Remington and Roldan Pozo. *NIST Sparse BLAS User's Guide*. National Institute of Standards and Technology.
13. Bjarne Stroustrup. *The C++ Programming Language.* Addison-Wesley, Reading, Massachusetts, second edition, 1991.
14. The HPC++ Working Group. HPC++ white papers. Technical report, Center for Research on Parallel Computation, 1995.

Design of a Data Class for Parallel Scientific Computing

Takashi Ohta

Japan Atomic Energy Research Institute

Abstract. We propose the design of a data class that offers SPMD parallelization facilities. The data class encapsulates all of the parallel procedures, so applications are written without concern for paralllization. The class interfaces are identical in the parallel and sequential cases, so a program can use either and run without change in a parallel or sequential environment. An example with the design applied to a CFD calculation is presented. Results show that good parallel efficiency is obtained by this approach.

1   Introduction

Parallel computing is now a practical candidate for a high-performance computing. However, in spite of its popularity, programming for parallel computing is still not an easy task. All the procedures for parallelization, such as domain decomposition, data distribution, data management, and data transfer must be written explicitly with a message passing library. Even with parallel compilers, insertion of directives that indicate what is to be parallelized is indispensable in order to obtain decent performance. Another difficulty is that a program written for sequential computation usually does not run on parallel computers without modifications. Often the program must be rewritten entirely for each parallel environment. It also costs extra debugging time to check if the algorithm performs correctly in parallel.

To write a parallel program, one must know both about a specific parallel environment and also parallel computation in general. Parallelizing an existing sequential code requires a detailed knowledge of the algorithm and the program, in order to know where and what is to be parallelized. This makes it difficult for an expert in parallel computing, but not in the application area, to rewrite a code.

These difficulties could be reduced if parallelization procedures and numerical methods were written separately. We propose a design that provides such a separation by encapsulating all the data and procedures related to parallelization in a class with variables for a scientific calculation. The data class allows access only to its variables but hides the parallelization procedures. With such a data class, a scientific computing code can be written using variables in the data class without regard to parallelism. Further, parallelization procedures only deal with class variables and its own data structures, regardless the application's algorithm.

2 Encapsulating Parallel Procedures

When writing a sequential code, variables are declared and a numerical method is written using these variables. If parallel programs could be written in the same way (using a class for variables and writing a program by using variables defined in that class), the burden in writing a parallel code would be reduced. Such a mechanism can be devised by designing a class with an appropriate access control, particularly in the data-parallel case, such as in domain decomposition. The data class has all the variables needed for a scientific application program. Figure.1 shows such a mechanism of a data class.

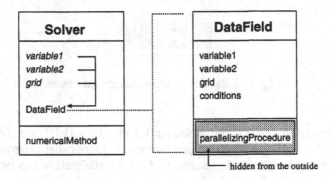

Fig. 1. Design for a data class.

All the data and procedures related to parallelization are also in the class. Since parallelization by domain decomposition depends only on the data structures and not with a numerical method, parallelization can be completed within the data class, and be hidden from the outside. The variables, however, are freely accessed and appear to reside in a class containing only data. With this design, the numerical method and parallelization procedures are separated as they are placed outside and within the class. A numerical method can be written as a sequential program by using the variables. Parallelization is done strictly within the class, using class variables, without knowing details of the numerical algorithm.

3 Unifying Interfaces

Hiding parallelization procedures within a class is not enough. These procedures must be invoked by calling them or sending them messages. The most striking difference between a parallel code and a sequential one is the need for data transfer among the parallel processes, arising from domain decomposition as used here. Two different programs would still need to be prepared for parallel and sequential computation.

To make a single program valid in both environments, we propose to unify the interfaces defined by the class. Figure 2 shows interfaces to each variable, numerical algorithm, and parallelizing procedure in the data class. In this case,

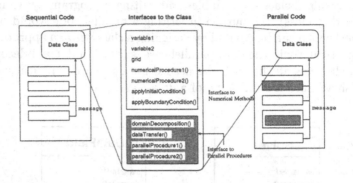

Fig. 2. A data class With specific interfaces.

a parallel program (right) and a sequential program (left) are different since there are calls to the parallelizing procedures in the parallel program. However, procedures for the numerical methods and for parallelization can be unified and given a common interface to both parallel and sequential programs as shown in Fig. 3. Both program can be written in the same way by using that common interface.

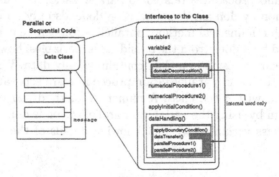

Fig. 3. A data class With unified interfaces.

When parallelization is done by domain decomposition, data transfer at boundaries of each decomposed sub-domain can be unified with the physical boundary conditions of the problem with a procedure that handles boundary conditions. Both parallel and sequential programs call the procedure regardless of the differences in internal processing. Since the parallelization procedures are hidden from numerical methods, the differences among various parallel environments are also isolated there, and numerical methods are not affected.

Encapsulating parallel procedures (the preceding section) and unifying interfaces are the keys to the design of data classes with the required characteristics.

4 Example: CFD

We apply the design to a CFD code. The example is a finite-volume solver of two-dimensional Euler equations. A structured grid is used. Roe's Riemann solver and the MUSCL method are used for estimating the numerical flux, and the explicit Euler's integration is used for time marching. Implementation is in C++.

The classes for the code are shown in Fig. 4. The class named FlowField

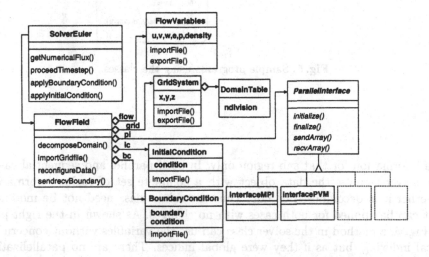

Fig. 4. A class diagram for a CFD solver.

is the data class, and includes classes for FlowVariables and GridSystem, the variables needed in CFD, as well as classes and procedures related to parallelization.

Programs can be written with these classes as shown in Fig. 5. The left program represents a sequential computation. The data class FlowField is instantiated and initialized. When the object is passed to the solver class SolverEuler, the instance contains data for the entire region.

For a parallel computation, the program is slightly modified as the right program shows: A declaration of an interface to the parallel environment is added, as is a call to decompose the region in the data object. This decomposition causes reconstruction of the input data inside the object. Therefore, when the data object is passed to the solver class, the instance has a decomposed sub-region and

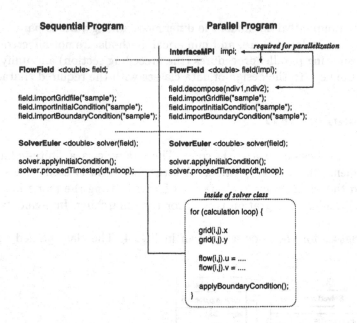

Fig. 5. Sample programs using the classes.

the information for that sub-region only. In both parallel and sequential cases, the solver receives the data object with a complete set of region information whether it is decomposed or not. Thus the solver class need not be modified, but can be applied for both cases with no changes. As shown in the right part of Fig. 5, a method in the solver class can use the variables without concern for local indexing, but as if they were global indices. There are no parallelization procedures explicitly inserted in the methods.

The resulting parallel performance is shown in Fig. 6. The calculations were carried out for two different grid sizes, 400x300 (120,000 points) and 600x500 (300,000 points). In both cases the speedup ratio is nearly linear through 48 nodes. Two factors contribute to this good parallel efficiency. First, the entire solver is implemented as a class and applied as a whole, achieving a large granularity of parallelism. Second is the treatment of boundary data transfers. Handling physical boundary conditions and internal decomposition boundaries separately would introduce special cases. For example, some sub-regions have two physical boundaries and two internal decomposition boundaries, while others have four internal decomposition boundaries only. The latter case would have to wait until the former had completed. In our design, decomposition data transfer and physical boundary condition are handled in a uniform way by the same method, simplifying processing.

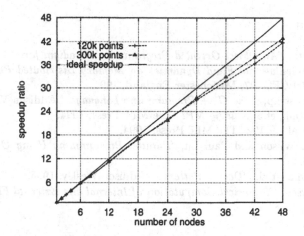

Fig. 6. Performance on IBM's RS6000 SP

5 Concluding Remarks

A data class design that supports parallelization internally and independently is proposed for writing parallel scientific programs. A code designed in this paradigm shows good parallel performance on IBM's RS6000 SP. Portability should be tested with other parallel machines in the future.

By encapsulating parallelizing procedures and unifying the interfaces to the class, a code can be executable in both parallel and sequential environments. While writing the code one is not concerned about parallelization, and development can proceed effectively in an interactive workstation environment. Once the code is completed it can run without change on large-scale parallel computers.

In an independent way, parallelization procedures can be implemented by dealing only with data, and not with particular scientific algorithms, enabling independent development of numerical methods and parallelization procedures. Many scientific algorithms use the same data structures, but manipulate them in different ways. Since parallelization procedures such as domain decomposition work with data, many data classes can be shared by different scientific applications. If such classes are prepared for various data structures (structured and unstructured grids, and particles, for example) writing parallel scientific codes can be made easier.

The ideas presented here are not limited in their applicability to specific libraries or numerical methods, but are a paradigm for building parallel programs. Although C++ is used for the implementation in this paper, other languages such as Fortran90 or Java can also be used in this way.

References

1. Takashi Ohta, *"An Object-Oriented Programming Paradigm for Parallel Computational Fluid Dynamics on Memory Distributed Parallel Computers,"* Parallel CFD '97, Manchester England, 1997.
2. Bjarne Stroustrup, *"The C++ Programming Language,"* Addison-Welsley, 1991.
3. William Gropp, et.al., *'using MPI'* The MIT Press, 1994.
4. Al Geist, et.al., *'PVM'* The MIT Press, 1994.
5. Gregory V. Wilson and Paul Lu, *'Parallel Programming Using C++'*, The MIT Press, 1996.
6. Erich Gamma, et.al., *'Design Patterns'*, Addison-Wesley, 1995.
7. Charles Hirsch, *'Numerical Computation of Internal and External Flows'*, John Wiley & Sons, 1988.

Describing Objects in Parallel
ECEM Image Reconstruction

Wei-Min Jeng

Department of Computer Science

University of Houston

Houston, Texas 77204

pet@cs.uh.edu

Camilliam Lin

Anadrill

Schlumberger Oil and Well Service

Sugar Land, Texas 77478

clin@sugar-land.anadrill.slb.com

Abstract

Early Communication Expectation Maximization (ECEM), one of the parallel image reconstruction techniques [1, 2, 3, 4], introduction was intended to speeding up the generation process of Positron Emission Tomography (PET) images. Despite the fact that it has been shown to be effective running on distributed memory parallel machine like IBM SP2 parallel machine, implementation independent issues still remain untouched in designing a better software architecture. In this study, we will characterize the properties of the ECEM application algorithm and try to find the objects that best describe the data structures of our particular scientific application. Abstract data typing techniques will be employed for specifying the framework of the application in both complete and precise manner. Message Passing Interface (MPI) primitives also will be embedded into the object model as our target programming paradigm. Parallel I/O activities also are performed using derived data types with the support of the underlying parallel file system. Being specified not by its object oriented language implementation, the design is used to perform the experiments on IBM SP2 parallel machine with MPI parallel library functions.

1 Early Communication Expectation Maximization (ECEM) Algorithm

PET machine, a system of detectors, captures the double photons emitting from the radiopharmaceuticals within the tissues of patient's body. The algorithm proposed by Shepp and Vardi [1, 2] has shown its proof in superior accuracy of reconstructed images. The algorithm proposed can be written as follows

$$\lambda^{new}(b) = \lambda^{old}(b) \sum_{d=1}^{D} \frac{n(d)p(b,d)}{\sum_{b'=1}^{B} \lambda^{old}(b')p(b',d)}$$

To obtain the best approximation with maximum likelihood, the above operation is performed iteratively before its convergence. In its equivalent form, each iteration of the algorithm can be further decomposed into the following four steps [3]:

1) $x = P^T \cdot \lambda,$
2) $\Phi_d = n_d / x_d,$ for d = 1, 2, ...D,
3) $y = P \cdot \Phi,$ and
4) 4) $\lambda_b^{new} = \lambda_b^{old} \cdot y_b,$ for b=1, 2, ...B.

Here P is a matrix where P(b, d) represents the probability of photons emitted from box b will be detected in detector pair d. n is a vector containing the number of photons detected by each pair. The goal is to find λ which represent the number of photons emitted from each box with best performance.

The parallel algorithm designed by Jeng and Huang [4] can be described as follows. Similar to the partition by detector scheme [5], we divide the P matrix (BxD) along the detector pair dimension into groups whose number equals the number of nodes. Instead of going ahead to complete the operations described in Step 1, we employ the independent property of the theoretical counts computation among detector pairs in each group and try to create another level of parallelism for the tasks on each processing node. Steps 1 to 3 are merged into one and partial results of y are sent over the network using total exchange reduction operations while remaining computation can be performed concurrently. Intended to overlap the communication with computation, the early communication can be performed by dividing the counts profile information and dispatching the sub-messages in batches on the fly once they are generated. It can be shown that the performance can be improved by using this new partition scheme with proper selection of early communication pattern.

2 Object Model

2.1 Abstraction of Objects

According to the ECEM algorithm, the approximation process of the activities for the target images can be executed in parallel with minimal communication overhead. It is important to identify the existing dependencies in the scheme in order to find the underlying properties for all components. As described in Section 1, the four separate steps each has a precondition before it can carry out its operations. Semantic properties are not elaborated here for specification of corresponding restrictions. The abstraction of each iteration can be formally specified using the following data structure excluding the effects of special parallel environment implementation. The objects can be classified into two categories, distributed objects and replicated ones, according to the specific parallelization scheme. For instance, the geometry information matrix will be partitioned into local nodes thus it falls into the type of distributed object. It is important to define two important steps of EM algorithm as distinct objects while optimizing the performance with hiding of communication latency.

Despite the fact that the experiments are performed on IBM SP2 machine with MPI software library primitives, the following descriptions of the classes and their objects can still be implemented for other configurations with reasonable efforts. Neither physically partition nor hardware architectural information is the factor here in our object design. Precise descriptions of our data objects are shown in Figure 1 using abstraction for the purpose of illustration.

```
class EM_MPI {
    int iteration_num_;
    MPI_COMM comm_group_;
    int node_id_;
    public:
        EM_MPI (int iteration_num, MPI_COMM comm_group, int
                node_id);
}; //common base class

EM_MPI::EM_MPI (int iteration_num, MPI_COMM comm_group, int
                node_id)
{
        iteation_num_ = iteration_num;
        comm_group_ = comm_group;
        node_id_ = node_id;
}

class REPLICATED public EM_MPI {
```

```
    int count_;
    MPI_FLOAT *vector_;
    public:
    domain (int *node_list, MPI_FLOAT *vector, int count) {
        count_ = count;
        for (int i=0; i<count; i++)
                vector_[i] = vector[i];


    }
}; //base class for all objects replicated on each processing node

class DISTRIBUTED public EM_MPI {
    int count_;
    MPI_FLOAT *vector_;
    public:
    domain (MPI_FLOAT *vector, int count) {
        count_ = count;
        for (int i=0; i<count; i++)
                vector_[i] = vector[i];
    }
}; // base class for all objects distributed on distinct local node

class GEOMETRY_INFO:public DISTRIBUTED {
    public:
    geo_matrix (MPI_FLOAT *original_matrix, int num_of_nodes, int
                num_of_boxes, int num_of_detectors, MPI_FLOAT
                **local_matrix) {
    for (int i=0; i<num_of_nodes; i++)
        local_matrix[i] = original_matrix += (num_of_boxes *
            num_of_detectors / num_of_boxes);
    }
}; // partition the geometric probability information into distributed nodes

class EXPECTATION:public DISTRIBUTED {
    int row_, col_;
    MPI_FLOAT *n_;
    MPI_FLOAT *x_;
    MPI_FLOAT *phi_;
    public:
    estimated_count (MPI_FLOAT **dist_matrix, MPI_FLOAT *lamda,
                MPI_FLOAT *x, MPI_FLOAT *n, MPI_FLOAT *phi,
                int row, int col) {
        row_ = row;
        col_ = col;
        n_ = n;
```

```
        trans (MPI_FLOAT**dist_matrix, MPI_FLOAT **trans_matrix,
            int row_, int col_) {
        MPI_FLOAT *x_ = Matrix_Vector (trans_matrix, lamda, row,
                                            col);
        ae_ratio (MPI_FLOAT *x_, MPI_FLOAT *n_, MPI_FLOAT
                    *phi_) {
            for (int i=0; i<col; i++) {
                    phi_[i] = n_[i] / x_[i];
            }
        }
    }
}; // expectation phase of EM


class MAXIMIZATION:public DISTRIBUTED {
    MPI_FLOAT **dist_matrix_;
    MPI_FLOAT *phi_;
    MPI_FLOAT *rec_y-, *y_;
    MPI_COMM  comm_group_;
    public:
    total_exchange (MPI_FLOAT **dist_matrix, MPI_FLOAT *phi, int
                    row, int col, int EC, MPI_COMM comm_group) {
        dist_matrix_ = dist_matrix;
        phi_ = phi;
        comm_group_ = comm_group;
        for (int i=0; i<EC; i++) {
                MPI_FLOAT y[i] = Matrix_Vector (dist_matrix[i], phi);
                MPI_handle handle[i] = MPIallreduce (y[i], MPI_FLOAT
                    recv_y_[i], col/EC, MPI_FLOAT, MPI_SUM,
                        comm_group);
        }
        for (int i=0; i<EC; i++) {
                wait (handle[i]);
                y_[i] += rec_y_[i];
        }
    MPI_FLOAT *update { return y_; }
    }
}; // maximization phase of EM;  EC - number of early messages
```

Figure 1: Abstraction of base and sub-base classes of ECEM objects.

2.2 Message Passing Paradigm

Our ECEM scheme, as in the SPMD computation, performs the parallelization tasks on SP2 machine by exploiting message passing programming paradigm. The programmer has to take care of the all the details of the message passing activities using library of primitives. To alleviate the difficulties of writing an error-free message passing application, efforts have been made [6, 7] to abstract the message passing programming for better software construction. To facilitate the object oriented modeling of scientific application like EM algorithm, it is equally important to establish the object models for both communication activities and computational tasks of parallel algorithm.

As we pointed out earlier, the fitting of our ECEM implementation with MPI bindings is critical for performance considerations. After analyzing the algorithm and current MPI progress, we find it is necessary to support the following MPI features and embed them in our modeling:

- data typing: used to prevent unnecessary type casting or any other inconsistencies,
- communication constructs: handles to objects such as communicators, nonblocking processes, etc.,
- library calls: names and function prototypes like total exchange functions should be followed for general interface.

2.3 Parallel I/O

Space constraints, along with the time constraints, limit the chances for reconstruction problem being solved by today's serial computer. In order to obtain satisfactory image quality, the size of the problem data grows proportionally with the increase of either image resolution or number of system detector pairs. In order to reduce the time in performing the input and output operations, the I/O operations have to be done in parallel before the actual reconstruction process can start. MPI-IO has standard built-in functions which provide flexible data accesses that traditional non-concurrent way of file access can not offer. Depending on the original file layout and the user-defined pattern, different processors can have accesses to various locations of the shared file simultaneously under IBM SP2 parallel file system. Similar to the regular parallel operations being performed under the distributed message passing programming model, the standardized interfaces transform the parallel I/O operations into message passing activities to provide a uniform programming environment for parallel application developers.

Using derived data type, the basic unit of the file information can be represented in elementary data type and acts as the basis for the complex data type. The data can be transformed into designated new format for different user-defined data partitions in this fashion with the support of the parallel I/O primitives. Our experiments indicate that the parallel I/O activities can achieve linear speedups proportional to the number of processor nodes.

3 Summary and Future Work

EM, algorithm used to estimate the unknown parameters through iterations, requires parallel implementations to overcome the problems with both time and space. Proper abstraction is useful in helping both software development and reuse with quick changing technology. We have shown in this paper how to describe the objects of our parallel ECEM application on message passing environment. Base class and sub-base classes are illustrated to take into accounts both software and hardware architecture for better modeling. Computational steps along with communication aspects are analyzed altogether in the design process without the compromise of performance. As discussed in Section 2, derived data types can be used to handle various patterns of parallel I/O with user defined partition schemes. Not just part of message passing events, I/O activities are strongly related to most of the parallel implementations thus these too can be included in our model in the future. Other issues such as load balancing, programming paradigms other than message passing also can be incorporated into the building of a comprehensive object model.

References

[1] L. A. Shepp and Y. Vardi, "Maximum Likelihood Reconstruction for Emission Tomography," *IEEE Transactions on Medical Imaging*, 1, 1982, 113-122.

[2] Y. Vardi, L. A.Shepp, and L. Kaufman, "A Statistical Model for Positron Emission Tomography," *Journal of the American Statistical Association*, 80, 1985, 8-37.

[3] Linda Kaufman, "Implementing and Accelerating the EM Algorithm for Positron Emission Tomography," *IEEE Transactions on Medical Imaging*, MI-6, 1987, 37-50.

[4] Wei-Min Jeng and Stephen Huang, "Efficient Parallel EM Image Reconstruction Algorithm: An Early Communication Approach," In Proceedings of nternational Conference on Parallel and Distributed Computing and Networks, 1997.

[5] Chung-Ming Chen and Soo-Young Lee, "On Parallelizing the EM Algorithm for PET Image Reconstruction," *IEEE Transactions on Parallel and Distributed Systems*, 5(8), 1994.

[6] B. McCandless, J. Squyres, and A. Lumsdaine, "Object Oriented MPI (OOMPI): A Class Library for the Message Passing Interface," In Proceedings of Second MPI Developer's Conference, 1996.

[7] Hua Bi, "Towards Abstraction of Message Passing Programming," In Proceedings of Advances in Parallel and Distributed Computing, 1997.

Flow in Porous Media Using NAO
Finite Difference Classes

Michael E. Henderson[1] and Stephen L. Lyons[2]

[1] IBM Research, T.J. Watson Research Center
[2] Mobil Technology Company

1 Introduction

In this paper we describe a set of C++ classes, implemented using the interfaces defined in the NAO (Numerical Analysis Objects) class library [6,8], for creating and manipulating variable coefficient finite difference operators and nonlinear equations involving them. We will show how these operator classes are used to define the equations governing the flow of compressible fluid in porous media, and how the equations are solved. Finally, we will show preliminary results of a time-dependent simulation.

The equations governing isothermal compressible flow in porous media are the mass conservation equation and Darcy's law for momentum balance. In general, compressible fluids have multiple components and multiple phases, but for this presentation we will restrict the problem to single-component, single-phase compressible flow in two dimensions. The mass conservation equation and Darcy's law are combined into a single time-dependent nonlinear partial differential equation. After specifying an initial condition and appropriate boundary conditions, this equation is then solved for pressure.

The finite difference approximation of this equation uses cell-centered pressures, edge-centered velocities, upwinding of fluid properties, and conservative averaging of rock properties. These are not trivial difference formulas, so while the example we give is small, this is a good example of the use of objects, and NAO's geometry/function/operator object interfaces, to build and solve a complicated set of equations. One of the advantages is that the system of nonlinear operators, its derivative, and the assembled Jacobian matrix are constructed together.

2 The Equation Governing Fluid Flow in Porous Media

We will restrict the problem to the flow of a single-component, single-phase compressible fluid in two-dimensional porous media at constant depth. The governing equation is obtained by substituting Darcy's law [3] into the mass conservation equation [10]:

$$\frac{\partial}{\partial t}(\rho\phi) - \nabla \cdot (\lambda[\mathbf{K}] \cdot \nabla P) - q = 0 \tag{1}$$

where ρ is the molar density of the fluid, ϕ is the porosity of the rock, $\lambda = \rho/\mu$, μ is the fluid viscosity, [K] is the permeability tensor of the rock, P is the pressure, and q is a source/sink term with units of moles/volume/time. Fluid properties and porosity depend on pressure and are obtained from equations of state, $\rho(P)$, $\mu(P)$, and $\phi(P)$. This equation is solved for pressure after an initial pressure distribution is specified and boundary conditions are specified zero normal flux,

$$(\nabla P) \cdot \mathbf{n}|_{\text{boundary}} = 0 .\tag{2}$$

3 The Finite Difference Scheme

The mass balance equation is a conservation law, and the primary concern for the differencing is that it produces a difference equation which is also a conservation law. This means that the mass balance should be an equation stating that the total mass flux into a mesh cell balances the mass contributed by the source term and the amount of mass being removed or injected into the cell giving the discrete form on a two dimensional Cartesian grid with constant grid spacings:

$$h_x h_y \left(\frac{\partial}{\partial t}(\rho\phi) - q \right)\Big|_{(x,y)} + h_x \left(\mathbf{F} \cdot \mathbf{i}|_{(x+\frac{h_x}{2},y)} - \mathbf{F} \cdot \mathbf{i}|_{(x-\frac{h_x}{2},y)} \right)$$
$$+ h_y \left(\mathbf{F} \cdot \mathbf{j}|_{(x,y+\frac{h_y}{2})} - \mathbf{F} \cdot \mathbf{j}|_{(x,y-\frac{h_y}{2})} \right) = 0 .\tag{3}$$

where the molar flux $\mathbf{F} = (-\lambda[\mathbf{K}] \cdot \nabla P)$.

This much is standard; the refinements come from how to average the coefficients $\lambda[\mathbf{K}]$. Fluid properties (λ) are upwinded for stability and rock properties ([K]) are harmonically averaged for flux continuity [2, 4]. Backward Euler is used for the time derivative and Newton's method is used to solve the subsequent set of nonlinear algebraic equations. Details of how the components of F are computed will be given later in this paper in the context of their implementation using the NAO Class Library.

4 Numerical Analysis Objects

The NAO Class Library defines abstract interfaces for three main categories of objects used in numerical analysis. These are *geometry*, *functions*, and *operators*. The intent of NAO is to define an interface to numerical tools that is independent of the particular data structures used to represent the inputs and outputs of these tools.

In NAO, discrete geometries have an interface which provides access to all of the information associated with a set of component cellular complexes. Geometrical information is associated with vertices, and topological information with a set of cells (whose *faces* are cells of one lower dimension). The flow simulation uses staggered rectangular grids, which are implemented by storing origins and mesh counts for each component.

Functions are mappings from one geometrical domain to another. They have a domain and range (which are geometrical regions), and an interface that allows them to be evaluated at points in their domain. Discrete functions have a number of "values" which define them. These may be thought of as the coordinates of the function in a finite-dimensional function space. The discrete functions used in this simulation store a single number at each vertex of the domain (TableFunctions). For example, the velocity is defined as mapping a staggered rectangular grid with two components to the real line. The TableFunction stores the value of the x-component of velocity on the first component mesh of the domain, and the y-component on the second component.

Operators are mappings from one functionSpace to another. Operators have a domain and range, and can be *applied* to a function in the domain. This operation returns the corresponding function in the range. Operators may have derivatives, and FiniteDimensionalOperators have derivatives which can be represented as linear matrices.

5   Finite Difference Operators

NAO Finite Difference Operators (FDO's) are represented by a set of coefficients, together with fractional shifts. The full definition in terms of quantities in the data structure representing the operator is

$$FDO(x) = \Sigma_{i=i0}^{i1} \Sigma_{j=j0}^{j1} f_{ij}(x) E_x^{i+\sigma_x} E_y^{j+\sigma_y} \ . \tag{4}$$

FDO's are constructed by specifying the size of the stencil (i0,i1,j0,j1), stencil coefficients (f_{ij}), and the points on which the operator will produce function values (the range of the operator). It is possible to give the range implicitly, by specifying where values of functions to which the operator will apply are available (the domain of the operator), and computing the largest possible range.

A second method of constructing FDO's is to use shift operators

$$E_x^{-1}, \ E_x^{-1/2}, \ E_x^{1/2}, \ E_x, \ E_y^{-1}, \ E_y^{-1/2}, \ E_y^{1/2}, \ E_y, \tag{5}$$

and to build up more complicated operators by addition, subtraction, composition, and multiplication by a scalar or a function.

6   Nonlinear Operator Equations

We have created composite NAO functions and operators which are constructed from pairs of functions and operators together with an operation. The resulting function or operator implements the interface routines in terms of the interface routines of the two objects from which it is constructed, and operations appropriate to the particular operation.

Instead of providing explicit representation of the result of each binary operation in an expression (often impossible), this produces a tree of simple or composite objects which represents the expression. This is called to "deferred execution" or "lazy evaluation" [5, 9]. The nonlinear operations are available either as overloaded C++ operators, or through a subroutine which parses an expression stored in a string and then performs the requested operations.

The approach of using "expression templates" [7, 11] allows compiler optimization of the composite operations. There are a couple of things that make its use problematic in this context. One is that the fluid and rock properties are obtained from an external package, and C++ source is not available for inlining. The other is that shift operators like Ex are context-dependent (in a single expression they mean different things in different places).

7 Building the Equations

The flow simulation is performed by writing an operator equation whose solution is the pressure at the next time step. This operator equation is solved by extracting the derivative as a linear matrix, and applying the operator to a current guess, then using a linear equation solver to calculate a Newton correction.

The equation is built in three steps. The permeability matrix is averaged, the fluid properties are upwinded, and then the gradient and divergence difference operators are defined and the whole equation written.

8 Averaging the Permeabilities

The averaged permeabilities use the harmonically averaged entries:

$$\bar{K}_{xx}^{i+1/2,j} = \frac{2K_{xx}^{i,j}K_{xx}^{i+1,j}}{K_{xx}^{i,j}+K_{xx}^{i+1,j}} , \quad \bar{K}_{yy}^{i,j+1/2} = \frac{2K_{yy}^{i,j}K_{yy}^{i,j+1}}{K_{yy}^{i,j}+K_{yy}^{i,j+1}} ,$$

$$\bar{K}_{xy}^{i+1/2,j+1/2} = \frac{4K_{xy}^{i,j}K_{xy}^{i+1,j}K_{xy}^{i,j+1}K_{xy}^{i+1,j+1}}{K_{xy}^{i,j}+K_{xy}^{i+1,j}+K_{xy}^{i,j+1}+K_{xy}^{i+1,j+1}} ,$$

$$\bar{K}_{yx}^{i+1/2,j+1/2} = \frac{4K_{yx}^{i,j}K_{yx}^{i+1,j}K_{yx}^{i,j+1}K_{yx}^{i+1,j+1}}{K_{yx}^{i,j}+K_{yx}^{i+1,j}+K_{yx}^{i,j+1}+K_{yx}^{i+1,j+1}} . \tag{6}$$

The entries $\bar{K}_{xy}$ and $\bar{K}_{yx}$ are defined at the corners of the grid (pressure is at the center of the cell), and $\bar{K}_{xx}$ and $\bar{K}_{yy}$ at the midpoints of the vertical and horizontal edges respectively. The averaged permeability tensor is

$$\begin{bmatrix} \bar{K}_{xx}^{i+1/2,j} & KXY \\ KYX & \bar{K}_{yy}^{i,j+1/2} \end{bmatrix} \tag{7}$$

where

$$KXY = \tfrac{1}{4}(E_y^{1/2} + E_y^{-1/2})\bar{K}_{xy}^{i+1/2,j+1/2}(E_x^{1/2} + E_x^{-1/2}) ,$$
$$KYX = \tfrac{1}{4}(E_x^{1/2} + E_x^{-1/2})\bar{K}_{yx}^{i+1/2,j+1/2}(E_y^{1/2} + E_y^{-1/2}) . \tag{8}$$

The additional averaging in the definition of the tensor is necessary because the tensor operates on the flux vector, and so the entries must exist at the point the components of the flux are defined.

The code to create the averaged permeability $\bar{K}_{xx}$ is

```
xxAvg=NASCreateNonlinearOperator(FS,"2/(1/SqrtEx+1/SqrtExInv)");
HKxx=xxAvg(Kxx);
```

The NASCreateNonlinearOperator subroutine parses the expression, and uses the functionSpace FS as the domain to construct the "predefined" operators SqrtEx and SqrtExInv. Then the expression is evaluated. The second line of code performs the averaging.

The next step is assembling the tensor.

```
AvgK=NASCreateNonlinearOperator(
    "[[Kxx,.25*(SqrtEy+SqrtEyInv)@(Kxy*(SqrtEx+SqrtExInv))],
     [.25*(SqrtEx+SqrtExInv)@(Kyx*(SqrtEy+SqrtEyInv)),Kyy]]",
    "[Function Kxx,Function Kxy,Function Kyx,Function Kyy]",
    HKxx,HKxy,HKyx,HKyy);
```

The "@" operation is composition. The four averaged tensor entries are passed to the subroutine as functions. The types of the non-"predefined" identifiers are declared in the second string, and the values are passed in a variable-length argument list.

We use a subroutine and parsed strings, rather than overloaded operations, because it is convenient to infer the domains of the shift operators in the expressions. For example, the SqrtEy shift operator appears twice, and has different meanings when it appears. This is the inference that was mentioned previously. Since the first SqrtEy is composed with the product of a function and an operator, the function defines the domain of SqrtEy. The second SqrtEy is multiplied by a function, so its range is known. In this situation no domain is provided, since the domains and ranges of all the operators can be inferred.

9 Upwinding Fluid Properties

Fluid properties are upwinded using the pressure field from the previous time step, and ThresholdFunction's. An upwinded density is

$$\rho_{i+1/2,j} = \begin{cases} \rho_{i,j} & V_x^{i+1/2,j} > 0 \ , \\ \rho_{i+1,j} & otherwise \ . \end{cases} \tag{9}$$

Threshold functions are step functions. They are constructed using a reference function. The ThresholdFunction evaluates to zero if the reference function evaluates to a number less than zero, and to one otherwise.

If we construct two ThresholdFunction's, one based on the x-component of the gradient of the old pressure field (tX), and another using the y-component (tY), we can upwind the fluid properties (λ) using shift operators:

```
Lambda=NASCreateNonlinearOperator(
    "[[(tX*(SqrtEx-SqrtExInv)+SqrtExInv)@lambda],
     [(tY*(SqrtEy-SqrtEyInv)+SqrtEyInv)@lambda]]",
    "[Function tX,Function tY,Operator lambda]",
    tX,tY,lambda);
```

10 Building the Equation on the Interior

The operator equation will act on the function P, and produce a function which maps a subset (the interior) of the mesh on which the pressure P is defined into the reals. First we define the gradient:

```
grad=NASCreateTwoDimensionalLinearFiniteDifferenceOperator(
        P0.getFunctionSpace(),
        "[[(SqrtEx-SqrtExInv)/hx],[(SqrtEy-SqrtEyInv)/hy]]");
```

where P0 is the pressure distribution from the previous time step. The divergence is then:

```
div=NASCreateTwoDimensionalLinearFiniteDifferenceOperator(
        AvgK.getRange(),
        "[[(SqrtEx-SqrtExInv)/hx,(SqrtEy-SqrtEyInv)/hy]]");
```

The grad acts on functions like P0, and div acts on the result of applying the averaged permeability tensor to the gradient. The constructors figure out at which points these operators will produce results. Finally, the operator equation for the interior is:

```
backwardsEuler=(rho*phi-rho(P0)*phi(P0))/dt
                    -div(Lambda*AvgK(grad));
```

11 Assembling and Solving the System

The equation on the interior has now been constructed. Operators for the boundary conditions (which produce functions mapping the boundaries into the reals) are written in a similar way, except that the ranges of the operators are specified so that values are produced on the boundaries. In addition, the appropriate flux terms are dropped to implement the no-flux boundary conditions. The result is a list of finite-dimensional operator equations G. The constructor for the list allows reordering of the equation numbers, which is necessary to obtain banded matrices.

We now wish to find P for which the operator produces the zero function. To do this we use Newton's method, which solves $G(P) = 0$ using the iteration:

$$P_{n+1} = P_n + \Delta P_n ,$$
$$G_P(P_n)\Delta P_n + G(P_n) = 0 . \tag{10}$$

For the purposes of this paper we have used an LU decomposition (LAPACK [1]) to solve for the update. The code for Newton's method uses the interface to the FiniteDimensionalOperator. It first applies the operator to the current guess, gets the derivative of the operator at the current guess as a LinearMatrix, factors the matrix and backsolves, and then adds the update to the current guess.

12   Results and Conclusions

Results of solving equation (1) with (2) are shown in Figs. (1) and (2). These figures show pressure contours at specific time intervals during the integration. The initial condition is a constant pressure field of 5000. The domain is square with a sink that withdraws fluid at a constant rate located in the upper left corner. The numerical grid (a 10x10, exponentially stretched grid) has been aligned with the principal directions of the permeability so that the permeability tensor is diagonal. Furthermore the permeability is anisotropic with $K_{xx}/K_{yy} =$ 100. Because of this anisotropy the pressure contours are not quarter circles but are elliptic in shape near the sink. The greatest pressure gradients are along the y-axis since the permeability in the y-direction is 100 times less than that in the x-direction. The pressure decreases everywhere in time as a result of the constant fluid withdrawal.

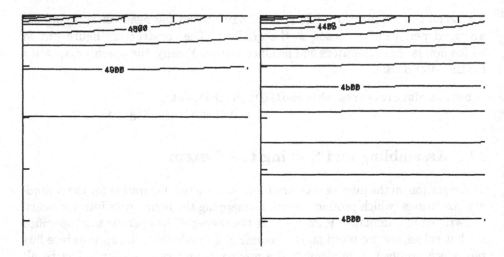

Fig. 1. a) Pressure after 4 days, min=4525.52, max=4999.46. b) Pressure after 28 days, min=4143.69, max=4951.51. Contour interval = 50, upper left quarter of domain.

As an approach, performing the whole simulation in C++, with operators built up via non-linear operations, is satisfying in that the program to build the operators resembles the numerical analysis. However, we have discovered that there are many assumptions that are made when such expressions are written. It is not easy to create the operators if these assumptions are not made by the software. For example, the user must declare the domain and range of each operator, and ensure that the operations are only performed on compatible objects. The constructors and reasonable defaults are critical.

The final performance of the code as a simulator leaves much to be desired. Techniques like expression templates and run-time optimization of expression trees may improve this. Another approach is to use the final operators to au-

232

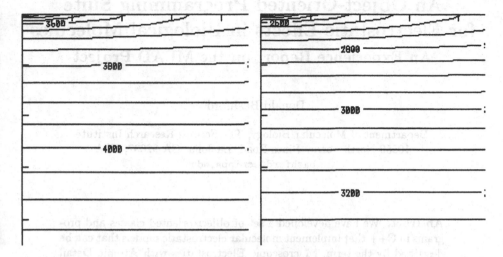

Fig. 2. a) Pressure after 124 days, min=3344.26, max=4429.90. b) Pressure after 252 days, min=2300.17, max=3595.85. Contour interval = 50, upper left quarter of domain.

tomatically write source code for applying the operators and assembling the Jacobians. That code can then be optimized by hand or by the compiler.

References

1. E. Anderson, Z. Bai, C. Bischof, J. Demmel, J. Dongarra, J. Du Croz, A. Green-baum, S. Hammarling, A. McKenney, and D. Sorensen. *LAPACK User's Guide*. Society for Industrial and Applied Mathematics, 1992.
2. Khalid Aziz and Antonin Settari. *Petroleum Reservoir Simulation*. Applied Science Publishers Ltd., 1979.
3. Henry Darcy. *Les Fontaines Publiques de la Ville de Dijon*. Victor Dalmont, Paris, 1856.
4. Bigyani Das, Stanly Steinberg, Susan Weber, and Steve Schaffer. Finite difference methods for modeling porous media flows. *Transport in Porous Media*, 17:171–200, 1994.
5. Robert B. Davies. Writing a matrix package in C++. In *Proceedings of the second annual object-oriented numerics conference*, pages 207–213, April 1994.
6. Craig C. Douglas, David A. George, and Michael E. Henderson. Object classes for numerical analysis. In *Proceedings of the second annual object-oriented numerics conference*, pages 32–49, April 1994.
7. Scott W. Haney. Beating the abstraction penalty in C++ using expression templates. *Computers in Physics*, 10(6):552–557, Nov/Dec 1996.
8. http://www.research.ibm.com/nao. *NAO Class Library*. IBM Research, 1996.
9. Scott Meyers. *More Effective C++*. Addison-Wesley, 1996.
10. Morris Muskat. *The Flow of Homogeneous Fluids Through Porous Media*. McGraw-Hill Book Co., Inc., 1937.
11. Todd Veldhuizen. Expression templates. *C++ Report*, 7(5):26–31, June 1995. Reprinted in C++ Gems, ed. Stanley Lippman.

An Object-Oriented Programming Suite for Electrostatic Effects in Biological Molecules

An Experience Report on the MEAD Project

Donald Bashford*

Department of Molecular Biology, The Scripps Research Institute
10550 North Torrey Pines Road, La Jolla, CA 92037, USA
bashford@scripps.edu

Abstract. We have developed a set of object-oriented classes and programs in C++ that implement molecular electrostatic models that can be described by the term, Macroscopic Electrostatics with Atomic Detail (MEAD). In the course of developing the MEAD suite, we have shifted from a class hierarchy rooted in atoms and molecules, to a system in which the top-level classes are the electrostatic potential and the entities that determine the potential in the equations of electrostatics: the charge distribution, the dielectric environment and the electrolyte environment. Atoms and molecules are then seen as objects giving rise to, or occurring as subclasses of, charge distributions, dielectric environments, etc. This shift in focus from the physical objects (molecules) to the more abstract objects that appear in the underlying physics has facilitated the development of alternative approximation schemes and numerical methods through subclassing. It also provides a natural way of writing high level programs in terms of potentials and distributions. Some of the newer elements of C++, such as templates and RTTI, have proven useful to solve multi-method and default method problems. MEAD is distributed as free software.

1 Introduction

1.1 Electrostatic Models

Electrostatic effects play a major role in determining the structure, function and chemistry of biological molecules; and the modulation of electrostatic effects by the aqueous and electrolytic environments that predominate in living cells is crucial [1]. Many simulations have been done in which both a biomolecule and several thousand surrounding water molecules are modeled explicitly at the atomic level. But for many purposes, such calculations are slow to converge and overly expensive. A macroscopic approach introduced by long ago by Born [2] has appealing simplicity and surprising accuracy for simple ions such as sodium

---

* The author gratefully acknowledges support from the National Institutes of Health (GM45607)

and chloride. The ion is modeled as a sphere of radius R containing a charge, Q. The region outside the sphere is considered to have a dielectric constant of 80 for water or 1 for vacuum. The differential work of charging for the solvated versus vacuum systems (and thus the electrostatic free energy of solvation) is then, $\Delta G = -Q^2 \left(1 - \frac{1}{80}\right)/2R$. The Born model gives quite reasonable agreement with measured solvation free energies of simple ions.

With the availability of atomic-level models of protein structures and high speed computers, it became possible to use numerical methods to incorporate the detailed shape of the molecule in the description of the dielectric boundary and charge distribution [3]. Meanwhile, similar ideas were being applied to the problem of solvent environments in small-molecule quantum chemistry calculations [4]. Such models are referred to here by the term, Macroscopic Electrostatics with Atomic Detail or MEAD, which is also the name for the C++ class and program suite described here. MEAD models have become very popular in the theoretical molecular biophysics community over the past 10 years or so [5, 6]. The present author's research group has been concerned with pH titration behavior of proteins [7]; titration and redox properties of small model compounds in conjunction with quantum chemical methods [8]; and protein and peptide stability [9]. The MEAD programming suite has been developed for, and used in these projects.

To state the MEAD model more explicitly, the solute is assumed to be an object of low dielectric constant (typically 1.0–4.0) with embedded charge, while the solvent (water) is modeled as a high dielectric continuum (typically 80). The locations of the embedded charges, the shape of the dielectric boundary and the boundary of the electrolytic region is determined by the coordinates and radii of the atoms making up the solute molecule. The electrostatic potential, $\phi(\mathbf{r})$ is usually governed by the linearized Poisson–Boltzmann equation,

$$\nabla\epsilon(\mathbf{r})\nabla\phi(\mathbf{r}) - \kappa^2(\mathbf{r})\epsilon(\mathbf{r})\phi(\mathbf{r}) = -4\pi\rho(\mathbf{r}) , \qquad (1)$$

where ρ is the charge distribution, ϵ is the dielectric constant which takes on different values in the molecular interior or exterior, and κ is a parameter that represents the effect of mobile ions in the solution. In MEAD, a finite difference method is typically used to solve this equation.

1.2 Goals

At its inception in 1990, the development of MEAD was an experiment in applying object-oriented techniques and the C++ language to theoretical molecular biophysics—a computationally intensive field that has been dominated by the procedural programming paradigm and the FORTRAN language. Other programs that use finite difference methods for protein electrostatics include the FORTRAN programs, DelPhi [10] and UHBD [11]. Recently, some multi-grid finite difference solvers have been implemented in C++ [12, 13] and applied to protein electrostatics. A key design goal of MEAD is to provide room for variation as to the methods of representing and solving the electrostatic problems that arise for molecular systems. For example, the electric charge generating the potential might be represented as a discrete set of point charges, q_i, or as a continuous density arising from a wavefunction calculated by a quantum chemistry program. Various numerical methods might be used to solve (1), or for simple geometries, an analytical solution might be used.

It is hoped that the MEAD programming suite will be useful to the molecular biophysics research community, particularly to those who are interested in developing new variations and applications of the MEAD model or related models. Therefore, MEAD is distributed under the terms of the GNU General Public License, and users are encouraged to study, modify and extend it to suite their own interests, and to communicate with the author so that generally useful extensions can be made available to all. MEAD can be obtained in both source-code and executable form from ftp://ftp.scripps.edu/pub/electrostatics.

2 The Natural Abstractions for the Problem

In the Poisson–Boltzmann equation (1) four kinds of entities participate—charge distribution, dielectric environment, electrolyte environment, and electrostatic potential—which are fairly general in electrostatic problems. These are the top-level classes in the analysis. More specific cases, such as point charges or anisotropic dielectrics, can be thought of in terms of more specialized subclasses. Charge distributions appear in two roles: They *generate* electrostatic potentials through relations such as (1). They also *feel* the electrostatic potential in the sense that there is an energy of interaction, such as,

$$U = \int_V \phi(r)\rho(r)d^3r \ , \tag{2}$$

between charge distributions and electrostatic potentials. The dielectric and electrolyte environments have a different kind of role: They respond to and thereby *modify* the electrostatic potential rather then generate it. Addition is a meaningful operation for both charge distributions and electrostatic potentials, and there can be multiplication between charges and potentials in the sense of (2). Typical calculations will begin by defining a dielectric and electrolyte environment and then setting up a number of charge distributions and calculating the potentials generated by each of them, perhaps exploiting the additivity properties of charges and potentials along the way.

Since MEAD is a system for *molecular* electrostatics, the analysis must involve molecules and their environments, such as solvents and membranes. In an earlier version of MEAD, the molecule was chosen as the main base class. This reflected the emphasis of traditional molecular mechanics packages. Electrostatic properties were then added as special attributes of molecules. This system became very awkward since the semantics of molecules had to change as different kinds of electrostatic models were implemented. It was therefore decided to rewrite the system in terms of the analysis outlined above. Molecules now enter the analysis through the electrostatic classes: A set of atomic coordinates and charges may be a *kind of* charge distributions. Atomic coordinates and radii, together with specifications of solvent properties, may *define* the dielectric and/or electrolyte environments. And molecules will generate or feel electrostatic potentials in their role as a kind of charge distribution.

3 Design and Implementation

3.1 Overview of MEAD Classes

The analysis of the problem domain suggests four top-level electrostatic classes: ChargeDist, DielectricEnvironment, ElectrolyteEnvironment and Elstat-Pot (for the electrostatic potential). We have implemented lower-level subclasses representing several situations: uniform dielectric and electrolyte, spherical geometry, dielectric slab (to represent membranes), and several classes in which the boundaries are determined by complex atom-level data. Charge distributions are generally collections of points, although we have made a preliminary implementation of charge densities represented as Slater-type functions. For the uniform, spherical and slab geometries, analytical or series solutions for the potential are implemented while for the atom-based geometries, numerical methods are provided. Equation (1) implies that the detailed data and methods of the Elstat-Pot will depend on the particular subclasses and data of the charge, dielectric and electrolyte objects that determine the potential. In MEAD, this is expressed through a constructor of ElstatPot that takes a DielectricEnvironment, a ChargeDist, and an ElectrolyteEnvironment as arguments. The constructors of the "input" classes such as DielectricEnvironment, require an explicit specification of the underlying subclass through a subclass pointer argument. The member functions of ElstatPot include value, field and displacement which return the electrostatic potential, the electric field, or the electric displacement, respectively, at a point in space. The member functions and operations of ChargeDist include a function to compute the total charge, and a "+" operator for adding charge distributions. The multiplication of a ChargeDist by an ElstatPot, as defined by (2) is also supported. Connection with molecular structures is made through the class AtomSet which stands outside the hierarchy described above, but is often used as a constructor argument to specify that a charge distribution or a dielectric boundary is determined by a set of atoms.

A small example will demonstrate how these classes are used in the higher levels of a program. Suppose one wishes to find the interaction energy of a superoxide ion with the potential due to a protein molecule. A program to do this could be written as follows:

```
float diel_in = 2.0, diel_wat = 80.0; ionic_str = 0.1;
AtomSet protein;
protein.read("protein_coords_charges_radii");
DielectricEnvironment eps_mol
    (new TwoValueDielectricByAtoms(protein, diel_in, diel_wat));
ElectrolyteEnvironment ely_mol
    (new ElectrolyteByAtoms(protein, ionic_str));
```

```
ChargeDist rho_mol(new AtomChargeSet(protein));
ElstatPot phi_mol(eps_mol, rho_mol, ely_mol);
phi_mol.solve();
AtomSet superoxide;
superoxide.read("superoxide_datafile");
ChargeDist ion_chg(new AtomChargeSet(superoxide));
cout << "interaction = " << phi_mol * ion_chg << endl;
```

3.2 Managing Polymorphism

The interdependent polymorphisms of the dielectric, electrolyte and charge, on the one hand, and the electrostatic potential on the other, require that the ElstatPot constructor, must "look inside" its arguments at run time to see what underlying subclasses are involved and then set up an electrostatic potential object that implements the appropriate methods and data structures. If there are d types of dielectric, c types of charges and e types of electrolytes, there could be as many as dce different construction pathways for the ElstatPot constructor to choose between. An effective means of dealing with this complexity is needed.

This problem is addressed in the context of the envelope–letter idiom [14] which is used for the top-level classes. An ElstatPot contains a pointer, rep, to an ElstatPot_lett object which contains the underlying data. The other top-level classes are similarly lightweight envelopes containing pointers to a "heavier" *_lett letter object. The ElstatPot constructor must query the rep pointers of its arguments and call the proper constructor of a subclass of ElstatPot_lett to assign to its own rep. Since C++ does not provide multi-methods [15] directly this is a somewhat awkward problem. At one point, it was solved by a double dispatch-like method [16] which required the dielectric subclasses to supply a virtual member function which called one of several required virtual member functions of the electrolyte, subclasses, and so on. This "triple dispatch" method was cumbersome and the addition of a new subclass of, say, electrolyte, could require the modification of the dielectric and charge subclass member functions. The introduction of run-time type identification (RTTI) into the C++ language and increasingly reliable support for templates made it possible to switch to a simpler table-lookup method.

The lookup table is a linked list of objects derived from the abstract base class, ElstatMaker, which contains the head of the list as a static pointer and provides a maker member function that uses RTTI to examine the subtype of its DielectricEnvironment_lett, ChargeDist_lett, and ElectrolyteEnvironment_lett arguments, and compares them to type information stored in the objects on the linked list. When a match is found, the derived_maker member function of the matching list item is called, which in turn calls the appropriate constructor for an electrostatic potential subclass. The maker function also allows for the provision of default construction methods through promotion of the input types to the more general *_lett base classes. When a new subtype is introduced into the system, for each relevant new combination of input types, a new subclass of ElstatMaker must be created, and an instance of it added to the static linked list. This task is eased by the provision of a template, DerivedElstatMaker that takes the input types and the output electrostatic potential type as arguments; and by the constructor of **ElstatMaker** which automates the building of the linked list. For example, if the introduction of subclasses with elliptical geometry required that the combination of an **EllipDiel** with an **EllipElyte** and any kind of charge distribution result in the construction of an electrostatic potential of class **EllipElePot** one would simply make the declaration,

```
DerivedElstatMaker<EllipElePot, EllipDiel,
                ChargeDist_lett, EllipElyte> elip_maker_ini;
```

at global scope (to that insure the list is set up before main starts).

A nearly complete implementation of the **ElstatMaker** class and the template is as follows:

```
class ElstatMaker {
public:
  ElstatMaker(const type_info& d, const type_info& c,
              const type_info& e) { // ... Add self to list
        next = list; list = this;
  }
  static ElstatPot_lett* maker(DielectricEnvironment_lett*,
            ChargeDist_lett*, ElectrolyteEnvironment_lett*);
private:
  virtual ElstatPot_lett* derived_maker(DielectricEnvironment_lett*,
    ChargeDist_lett*, ElectrolyteEnvironment_lett*) const = 0;
  ElstatMaker *next;
  static ElstatMaker* list;};

template<class Elstat_T, class Diel_T, class Charge_T, class Ely_T>
class DerivedElstatMaker : public ElstatMaker {
  // (constructor omitted)
  ElstatPot_lett* derived_maker(DielectricEnvironment_lett* dept,
    ChargeDist_lett* cdpt, ElectrolyteEnvironment_lett* eept) const
    {
      Diel_T* der_dept = dynamic_cast<Diel_T*>(dept);
      Charge_T* der_cdpt = dynamic_cast<Charge_T*>(cdpt);
      Ely_T* der_eept = dynamic_cast<Ely_T*>(eept);
      return new Elstat_T(der_dept, der_cdpt, der_eept);}};
```

The multi-method issue also arises in the implementation of operations such as the addition of two different charge distributions, or the multiplication of charge times potential. For these cases, the double dispatch or delegated polymorphism approach [15, 17] works reasonably well but RTTI can be used to reduce the impact of adding new classes.

3.3 Mandatory Methods

In order to provide reasonable default methods for operations such as obtaining the potential, it has been found necessary to require certain member functions of all classes within particular parts of the hierarchy, even when these functions are not entirely natural to all the subclasses concerned. For example, the provision of a default charge addition operation is made possible by requiring that all charge distributions provide expressions of themselves as a set of point charges, and methods for iterating over these points. This is quite natural for the Atom-ChargeSet class used in the example in Sec. 3.1. But it is not so natural for a class that represents a charge distribution as a set of Gaussian or Slater-type functions. Since the default method for finding the potential is a finite-difference calculation, all dielectric and electrolyte geometries, even those that could be solved analytically, are required to provide methods of mapping themselves onto a finite-difference lattice of specified parameters.

4 Application Programs

The MEAD suite includes several application programs that use the type system described above. A fairly simple example is solvate, which calculates the solvation energy of a molecule using the MEAD equivalent of the Born method. It begins like the example of Sec. 3.1, except that the potential, phi_mol, is multiplied by the charges that generate it, rho_mol, rather than by an outside set of charges. A second such calculation is made for the vacuum case, and half the difference is the electrostatic solvation energy. A much more complex example is multiflex, which implements a scheme for calculating the titration behavior of multiple sites in a protein in either a single conformer or a multi-conformer [7] case. The calculations involve setting up the dielectric and electrolyte environments of the protein, and then calculating the energetics of a series of local charge distributions representing the titrating sites in their protonated and unprotonated forms. Model compound calculations for the various sites are also performed.

5 Conclusions and Future Directions

Within the author's research group, MEAD has proven to be a powerful and flexible environment for creating programs to deal with molecular electrostatic problems, and to explore new methodological ideas. A crucial feature of the type system is the clean separation between the abstractions of electrostatic theory and the abstractions of molecular structure. Another important feature is that numerical methods are isolated in fairly low levels of the code, so that it should be possible to substitute different numerical methods without much effect on application code. At present however, only an SOR solution of the finite-difference equations is provided. The multiflex program has recently been parallelized using MPI at a coarse-grain level (the level of titrating sites). Opportunities for fine-grained parallelization of the sort commonly used in finite-difference methods exist at lower levels of the code. The current implementation makes use of the container classes provided by the GNU project's libg++, and its Pix style of iteration. It is planned to replace this with the Standard Template Library and its iterator model.

References

1. Perutz, M. F.: Electrostatic effects in proteins. Science **201** (1978) 1187–1191
2. Born, M.: Volumes and heats of hydration of ions. Z. Phys. **1** (1920) 45–48
3. Warwicker, J., Watson, H. C.: Calculation of the electric potential in the active site cleft due to α-helix dipoles. J. Mol. Biol. **157** (1982) 671–679
4. Miertus, S., Scrocco, E., Tomasi, J.: Electrostatic interaction of a solute with a continuum: A direct utilization of *ab initio* molecular potentials for the prevision of solvent effects. Chem. Phys. **55** (1981) 117–129
5. Sharp, K. A., Honig, B.: Electrostatic interactions in macromolecules: Theory and experiment. Annu. Rev. Biophys. Biophys. Chem. **19** (1990) 301–332
6. Honig, B., Nicholls, A.: Classical electrostatics in biology and chemistry. Science **268** (1995) 1144–1149
7. You, T., Bashford, D.: Conformation and hydrogen ion titration of proteins: A continuum electrostatic model with conformational flexibility. Biophys. J. **69** (1995) 1721–1733
8. Li, J., Fisher, C. L., Bashford, D., Noodleman, L.: Calculation of redox potentials and pK_a values of hydrated transition metal cations by a combined density functional and continuum dielectric theory. Inorg. Chem. **35** (1996) 4694–4702
9. Ösapay, K., Young, W. S., Bashford, D., Brooks, III, C. L., Case, D. A.: Dielectric continuum models for hydration effects on peptide conformational transitions. J. Phys. Chem. **100** (1996) 2698–2705
10. Nicholls, A., Honig, B.: A rapid finite difference algorithm, utilizing successive over-relaxation to solve the Poisson-Boltzmann equation. J. Comp. Chem. **12** (1991) 435–445
11. Madura, J. D., Briggs, J. M., Wade, R. C., Davis, M. E., Luty, B. A., Ilin, A., Antosiewicz, J., Gilson, M. K., Bagheri, B., Scott, L. R., McCammon, J. A.: Electrostatics and diffusion of molecules in solution: Simulations with the University of Houston Brownian Dynamics Program. Computer Physics Communications **91** (1995) 57–95
12. Oberoi, H., Allewell, N. M.: Multigrid solution of the nonlinear Poisson–Boltzmann equation and calculation of titration curves. Biophys. J. **65** (1993) 48–55
13. Holst, M., Kozack, R. E., Saied, F., Subramaniam, S.: Treatment of electrostatic effects in proteins: Multigrid-based newton iterative method for solution of the full nonlinear Poisson–Boltzmann equation. Proteins: Struc. Func. and Genet. **18** (1994) 231–245
14. Coplien, J. O.: Advanced C++ Programming Styles and Idioms. Addison-Wesly. 1992
15. Stroustrup, B.: The Design and Evolution of C++. Addison Wesley. 1994
16. Ingalls, D. H. H.: A simple technique for handling multiple polymorphism. in *Proc ACS OOPSLA Conference, Portland, OR.* 1986
17. Coplien, J. O.: Advanced C++ Programming Styles and Idioms. Addison-Wesly. 1992 pp. 134–140

A Portable, Object-Based Parallel Library and Layered Framework for Real-Time Radar Signal Processing

Cecelia DeLuca, Curtis W. Heisey, Robert A. Bond, and Jim M. Daly
cdeluca@ll.mit.edu, heisey@ll.mit.edu, rbond@ll.mit.edu

Massachusetts Institute of Technology Lincoln Laboratory*
244 Wood Street, Lexington, MA 02173-9108

Abstract. We have developed an object-based, layered framework and associated library in C for real-time radar applications. Object classes allow us to reuse code modules, and a layered framework enhances the portability of applications. The framework is divided into a machine-dependent kernel layer, a mathematical library layer, and an application layer. We meet performance requirements by highly optimizing the kernel layer, and by performing allocations and preparations for data transfers during a set-up time. Our initial application employs a space-time adaptive processing (STAP) algorithm and requires throughput on the order of 20 Gflop/s (sustained), with 1 s latency. We present performance results for a key portion of the STAP algorithm and discuss future work.

1 Introduction

For large-scale, real-time applications on parallel computers such as the Cray T3E and the Mercury RACE series, both performance and modularity are critical. We have developed an object-based, layered framework and an associated signal processing library for real-time radar applications. Initially the library will be used to perform space-time adaptive processing (STAP) algorithms, which are used in radar systems to suppress interference and detect targets. The algorithms consist of a large data cube transiting through a sequence of signal processing stages, or subsystems, and include routines such as matrix multiplies, matrix factorizations, and FFTs (Ward 1994). A typical operation at a given stage involves slicing the data cube into matrices and performing independent operations on the set (Figure 1). We have incorporated both task and data parallelism into our libraries so that multiple subsystems can be pipelined, and operations within a subsystem can be performed in parallel. Data redistribution occurs both within and between subsystems.

Our goals are portability, ease of use, and reusable software components along with the competing constraints of high throughput (20 Gflop/s sustained)

---

* This work was sponsored under Air Force Contract F19628-95-C-0002. Opinions, interpretations, conclusions, and recommendations are those of the authors and are not necessarily endorsed by the United States Air Force.

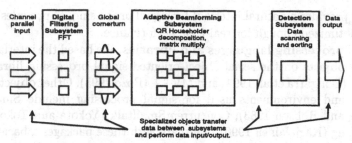

Fig. 1. The STAP application combines both task and data parallelism. Data (arrows) moves through a series of tasks (boxes), which may themselves be pipelines of tasks.

and low latency (about 1 s). We found no existing software environment which satisfied all of our requirements.

The object-oriented, C-based approach satisfies the demands of portability, modularity, and reusability and allows the use of a standard compiler. We improve portability through layering as well, by restricting machine-dependent code to a low-level communication kernel. To achieve real-time performance, we prepare for data transfers and perform allocations at set-up time. We have also created mechanisms for graceful departure from object-based design when performance requirements demand it, though initial performance comparisons suggest that this capability is rarely needed.

We begin by discussing related research. Next we provide an overview of our parallel programming model, which describes how classes relate to parallelism and flow of control. We describe the layers of our framework, reviewing the function and structure of each level, and provide detailed descriptions of the major classes. We conclude with a performance analysis of a key library routine and discuss our plans for future work. Standard definitions for architectural taxonomy and programming paradigms are used (e.g, Wilson, 1995).

2 Related Work

Our work lies at the intersection of several research areas. One area is that of parallel object-oriented languages for general purpose use; another is the development of serial and parallel object-oriented languages and libraries just for signal processing. We also integrate into our work research on optimal mappings for algorithms such as STAP on parallel machines. Below, we highlight the work in each of these fields which is most closely related to our own.

General-purpose object-oriented libraries and languages. A multitude of general purpose parallel, object-oriented languages, libraries and frameworks have been developed over the past decade, including C++// (Caromel et al. 1996), and POOMA (Reynders et al. 1996). We share basic design constructs with many of these packages. The layers of our software are simpler than but close to the multi-tiered structure of POOMA. Like the active objects in C++//, our software includes data objects which coordinate their own computations, data transfers and synchronizations. However, none of the environments we have encountered provides a natural interface for signal processing, or provides methods that are optimized enough for real-time performance.

Signal processing languages and libraries. We based the interface to our local data objects on the serial object-oriented signal processing libraries LAPACK++ (Dongarra et al. 1994) and MV++ (Pozo 1995). Other object-oriented languages and environments used for signal processing include Smalltalk-80 (Goldberg and Robson 1983), ConcurrentSmalltalk (Yokote and Tokoro 1995), and QuickSig (Karjalainen 1990). We share with these packages a basic strategy of making functionally related signal processing routines into classes. However, they are not in general designed for use on massively parallel processors or for real-time applications.

STAP and related implementations. A number of studies (Hwang and Xu 1996, McMahon and Teitelbaum 1996, Hwang et al. 1996) demonstrate that the real-time requirements for STAP and similar algorithms can be met with the current generation of general purpose parallel computers, using combinations of data and task parallelism.

3   Parallel Programming Paradigm

Though our application is SPMD, we support a hybrid of task and data parallelism. After an initial set-up, the application branches into multiple subsystems which provide task parallelism at a coarse-grain level. Subsystems in turn may contain combinations of data parallel operations and internal pipelines. Tasks within subsystems may be round-robin'd between node sets for routines which do not parallelize well but have high throughput requirements.

Subsystems process data as it becomes available, and otherwise exist in an idle state. When a subsystem finishes processing data, it forwards its results via a specialized data transfer object to the next subsystem in the signal processing chain. Our coarse-grain interactions are therefore inspired by a data flow approach, in that data dependencies and availability dictate task scheduling.

Classes of objects provide an encapsulation of coherent sets of data and methods to operate on the data. Because of the hierarchical layering of objects in our framework, an object marshals the activity of its constituent objects. For example, a subsystem might initiate data parallel Householder decompositions on a set of distributed matrices. Each of the matrices would organize its own data movements and synchronizations. By the time the node level was reached, all the data that a node would need to perform a computation would have been moved into place. The lowest level in the data class hierarchy is a node-level object, which simply operates on local data.

4   Layered Framework

The layered framework consists of an application layer, a mathematical library layer, and a machine-dependent kernel layer. At the application and library levels, we have created classes that internally coordinate the details of distribution, data transfers, and synchronization. In the kernel we have extended the basic C language types to include classes of mathematical objects, such as matrices and vectors, which are local to a node. Part of the kernel level is not structured using classes, but as a set of functions that may be called from any tier of the code.

The *application layer* is essentially the translation of the STAP algorithm to code, using C with library extensions. Application code is organized using object-based principles, with each major signal processing stage designed as a subsystem object. A digital filtering subsystem class, for example, is created with a set of methods which includes pulse compression and equalization.

The *mathematical library* consists of distributed matrix, vector and scalar data classes and their associated mathematical methods. The library interface to the application level is designed so that the application programmer views mathematical and engineering constructs, and not the details of data object distribution, data transfers, or optimizations.

A large percentage of the computation in STAP occurs in a few key routines, such as matrix multiply and QR Householder decomposition. The libraries themselves are written using data classes, except for these critical methods, which are optimized with low-level code. Throughout the code and particularly in inner loops, we often use macros for accessing class attributes and for indexing.

The *kernel layer* provides a uniform interface for machine-dependent functions. It consists of a model of the underlying machine architecture, platform-specific computation primitives and platform-specific communication functions. Functions performed by the kernel include synchronizations and data broadcasts across sets of nodes. We intend to use the same kernel source code for each platform and application, with compiler directives to select the appropriate code segments.

Our layered, object-oriented framework allows experts in radar signal processing to develop application code while parallel systems analysts decide how data objects should be distributed for a given machine size and architecture. It also enables highly optimized versions of mathematical methods and communication routines to be developed simultaneously by experts in parallel programming.

This approach has allowed us to construct and test our system systematically and rapidly, satisfying our ultimate constraints of reliability and delivery deadlines. It is also intended to minimize the time and effort spent modifying our code for new applications and for porting to different platforms.

5 Class Design

The primary abstract classes which constitute our framework are the Virtual Machine class and the Basic Type data classes at the kernel layer; Map, Atlas and the Mobile Type data classes at the mathematical library layer; and Subsystem and Conduit classes at the application layer (Figure 2).

Kernel Layer. A *Virtual Machine* class insulates application programmers and software engineers from the physical topology of a parallel machine. Subsets of

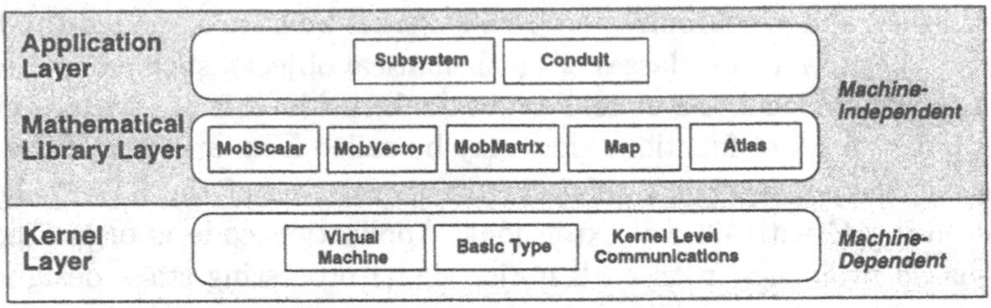

Fig. 2. Class design and layered framework.

nodes are viewed as Cartesian grids of three or fewer dimensions, a virtual topology which is suitable for radar signal processing algorithms. Generally, physical nodes are mapped one-to-one to nodes of a Virtual Machine, though this is not required.

Subsystems and data objects in the library layer each contain a Virtual Machine which defines on which nodes the object will have local data or a local component. Each Virtual Machine has its own Cartesian coordinate system, so that general routines can be written for differently sized and distributed data objects. Data objects which may be distributed over Virtual Machines include matrices and vectors, and arrays of matrices, vectors and scalars. We require that the Virtual Machine over which a data object is distributed be of the same or fewer dimensions than the data object, and restrict the distribution of a single mathematical dimension to a single Virtual Machine dimension. Also, we permit only block-cyclic distributions of matrices and vectors over Virtual Machines and replicated Virtual Machines. This enables us to use a general data redistribution scheme, discussed in more detail later.

We utilize the notion of processor sets to simplify coding and improve performance of matrix and vector algorithms. A processor set containing all the nodes in a Virtual Machine is built when a data object is constructed. Processor sets may also be built for the rows and columns of a Virtual Machine.

The *Basic Type* data classes includes scalars, vectors, and matrices which are local to a node. Basic Type objects do not have the capability to participate in interprocessor communication. Their attributes are minimal and include the size of the object. The methods of a Basic Type include standard mathematical operations and may be implemented using platform-specific, optimized routines.

Mathematical Library Layer. The *Map* class contains the Virtual Machine class and a Distribution Description class. The role of Virtual Machines is described in the preceding section; the Distribution Description class specifies parameters such as blocksize for each dimension. Maps are associated with a particular arrangement of data on a Virtual Machine and not a specific data type or size. Map methods include routines for conversion between mathematical and local indices.

All of the Maps in an application are stored in an *Atlas* class which may be read in from a file at set-up time. Thus the entire application can be remapped without rewriting or recompiling code. The Atlas greatly facilitates scalability

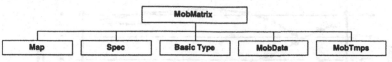

Fig. 3. Object diagram of a Mobile matrix

in that the dimensions of data objects can be changed, or the application run on differently sized machines, simply by modifying the contents of the Atlas. The application mapping may also be done automatically by an external program which analyzes the algorithm and machine architecture, though we do not yet support such a feature.

Data classes for mathematical arrays on parallel platforms differ from those created for a serial machine in that they may need to integrate into their design mechanisms for distribution over multiple processors and/or movement between processors. We refer to the ability to participate in interprocessor communication as mobility. Since distribution and mobility tend to be coupled, we have initially incorporated capabilities for these attributes into a single set of *Mobile Type* data classes. Mobile objects coordinate their activities through internal structures which execute data transfers and synchronizations. The Mobile classes we have created include scalars, vectors, matrices, and arrays of all of these.

A simplified object diagram of a Mobile matrix is shown in Figure 3. The primary components of the class are a Map, a Spec, a MobData, a Basic Type and a MobTmps class. The role of the Map has already been discussed. The Spec contains frequently used parameters which are convenient to store, such as the number of elements allocated per processor. The MobData class houses the machine-dependent parameters used for interprocessor communication. The data local to a node in a Mobile, distributed object is referenced to the data in a Basic Type. The Basic Type structure, described earlier, contains the parameters, such as rows and columns, that pertain to the local data. Thus for routines on data blocks that don't require interprocessor communication, a Basic Type method can be called from within a Mobile Type method. This has the great advantage of keeping the code for Mobile, distributed objects largely free of machine dependent calls, while still being efficient. Finally, the MobTmps class contains a variety of objects which are used for method optimization, array manipulations, and data transfers. The data transfer mechanism is described in more detail below.

Mobile objects use an object-based implementation of the PITFALLS scheme (Ramaswamy 1996) to coordinate redistribution of arbitrarily dimensioned block-cyclic arrays on source and destination processor sets that may be differently sized. We have algorithmically extended the PITFALLS scheme so that subsets of block-cyclic arrays can be moved efficiently as well. Routing information is stored in lookup tables so it does not have to be computed at run-time. These lookup tables and other supporting structures, such as staging buffers for packing and unpacking, are stored in MobTmps.

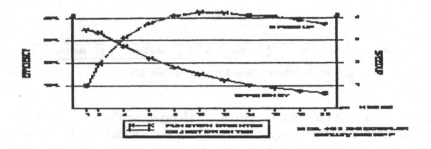

Fig. 4. QR Householder performance.

Application Layer

A *Subsystem* superclass has subclasses specialized for each of the major signal processing functions of our application (e.g. digital filtering, adaptive beamforming, and detection). The class enables high-level control mechanisms to treat clusters of nodes performing complex but functionally related operations as simple, unified objects. Inheritance is used to provide common methods and a consistent interface to all of the specialized Subsystems.

Conduits transfer data between Subsystems. A Conduit object is constructed using the Maps of the data objects that it picks up and delivers. Each Conduit is specialized for a particular data type and for a particular transfer. Conduits may replicate data, redistribute data in a different block-cyclic manner, or cornerturn data before delivering it. Conduits may store structures at set-up time for later use at run-time, and can manage multiple buffering of data so that communications between Subsystems can be overlapped with computation.

6 Performance Assessment

We present the results of the QR Householder transformation, as it represents a typical parallel operation and accounts for a sizeable portion of the algorithm workload. Figure 4 shows the performance of the Householder method (courtesy Arakawa, 1997). The Arakawa QR Householder study shows that a matrix distributed over a single row of processors provides the optimal mapping. An efficiency of 30% to 25% is achieved for distribution over four to six processors. The application requires an efficiency of 25% or greater in order for the subsystem to meet its allotted latency.

As indicated by the graph, there is no notable performance difference between the object-based version and the procedural versions. In the object-based version, we have allowed optimizations such as replacing indexing methods with macros and the breaking of encapsulation in innermost loops by accessing object attributes directly. The additional layers of function calls imposed by the object design do not result in a significant degradation of performance.

7 Future work

The current library is implemented in C and runs on a Mercury RACE platform. We are currently porting the library to our in-house Cray T3E-900 and producing a C++ version of our code. Future work will include optimization and performance evaluation of both the application and the library as a whole.

With our library and framework, we merge object-oriented programming, digital real-time signal processing, and massively parallel computing. We retain the benefits of an object-oriented approach yet achieve the performance required by a real-time system.

References

Arakawa, M.: "Analysis, Implementation and Performance of a Distributed Householder QR Factorization", *First Annual Workshop on High-Performance Embedded Computing*, HPEC '97, MIT/Lincoln Laboratory, Sept 17-18, 1997.

Bond, R.: *Measuring Performance and Scalability Using Extended Versions of the STAP Processor Benchmarks*, MIT Lincoln Laboratory, Technical Report, 1994.

Caromel, D., F. Belloncle, and Y. Roudier: " C++//,' 'in *Parallel Programming Using C++*, Ed., G. V. Wilson and P. Lu, MIT Press, 1996.

Dongarra, J. J. et al.: *LAPACK++: A Design Overview of Object-Oriented Extensions for High-Performance Linear Algebra*, Oak Ridge National Laboratory, University of Tennessee, Apr 1994.

Goldberg, A. and D. Robson: *Smalltalk-80: The Language and its Implementation*, Addison-Wesley, 1993.

Hwang K., and Z. Xu: "Scalable Parallel Computers for Real-Time Signal Processing", *IEEE Signal Processing Magazine*, Jul 1996.

Hwang K., Z. Xu and M. Arakawa: "Benchmark Evaluation of the IBM SP2 for Parallel Signal Processing", *IEEE Transactions on Parallel and Distributed Systems*, May 1996.

Karjalainen, M.: "DSP Software Integration by Object-Oriented Programming: A Case Study of QuickSig", *IEEE ASSP*, 7:2, Apr 1990.

McMahon, J., and K. Teitelbaum: "Space-Time Adaptive Processing on the Mesh Synchronous Processor", *IEEE IPPS '96* Honolulu, HW, USA; Apr 15-19, 1996

Pozo, R.: *MV++ v.1.5 - Matrix/Vector Class Reference Guide*, Applied and Computational Mathematics Division NIST, Oct 1995.

Ramaswamy, S., B. Simmons, and P. Banerjee: "Optimizations for Efficient Array Redistribution on Distributed Memory Multicomputers," *Journal of Parallel and Distributed Computing*, 38(2), Nov 1996.

Reynders, J. V., P. J. Hinker, J. Cummings, S. R. Atlas, S. Banerjee, W. F. Humphrey, S. R. Karmesin, K. Keahey, M. Srikant, and M. Tholburn: "POOMA," in *Parallel Programming Using C++*, Ed., G. V. Wilson and P. Lu, MIT Press, 1996.

Ward, J.: *Space-Time Adaptive Processing for Airborne Radar*, MIT Lincoln Laboratory, Technical Report 1015, Dec 13, 1994.

Wilson, G. V.: *Practical Parallel Programming*, MIT Press, 1995.

Yokote, Y and M. Tokoro: "Concurrent Programming in Concurrent Smalltalk", in *Object Oriented Concurrent Programming*, ed. A. Yonezawa and M. Tokoro, 1994.

Aspect-Oriented Programming of Sparse Matrix Code

John Irwin, Jean-Marc Loingtier,
John R. Gilbert^ , Gregor Kiczales, John Lamping,
Anurag Mendhekar, Tatiana Shpeisman
Xerox Palo Alto Research Center*

The expressiveness conferred by high-level and object-oriented languages is often impaired by concerns that cross-cut a program's basic functionality. Execution time, data representation, and numerical stability are three such concerns that are of great interest to numerical analysts. Using aspect-oriented programming we have created AML, a system for sparse matrix computation that deals with these concerns separately and explicitly while preserving the expressiveness of the original functional language. The resulting code maintains the efficiency of highly tuned low-level code, yet is ten times shorter.

Introduction

Traditionally there are two ways of writing sparse matrix code. One is to write in a high-level matrix language that offers fast prototyping and easily readable code. Unfortunately the result is usually too slow to use for real problems. The problem is that efficient sparse matrix code requires careful choice of data structures depending on the structure of the computation. The data structures must store only the non-zeros, and do so in a way that fits well with the computation's needs. Further, the access patterns of the computation must be exploited to achieve efficient access and update while allowing the computation to avoid operations on zeros.

The other approach is to write in a low-level language such as C or Fortran, directly implementing and exploiting the most appropriate data structures.

---

^ The work of this author was partially supported by DARPA under contract DABT63-95-C0087.

* 3333 Coyote Hill Road, Palo Alto, CA 94304, USA. gregor@parc.xerox.com,
lamping@parc.xerox.com

While this approach can give good performance, such a program generally takes much longer to write, and is less readable and less maintainable. Neither alternative is attractive.

We used the aspect-oriented programming (AOP) approach [1] [2] to resolve this dilemma. We built a language environment called AML which lets the user write high-level sparse matrix code along with annotations describing an efficient implementation. We picked a set of representative sparse matrix algorithms to use as test cases; one such algorithm is described in detail in this paper. As efficiency is a quantitative result, we measured the speed of our code compared to standard versions of the same algorithm. The result was quite satisfactory – our code matches the speed of the standard version based on the same algorithm, yet is considerably shorter and less complex.

An example: sparse LU factorization

We will demonstrate our approach on a typical numerical analysis algorithm: LU factorization by Gaussian elimination with partial pivoting (GEPP). An engineer would normally invoke sparse GEPP from a library rather than coding it by hand; in that sense this example is artificial. However, GEPP is representative of a large and important class of algorithms, namely incomplete factorization preconditioners [3], that are often coded by hand for domain-specific numerical modeling applications.

The basic algorithm for Gaussian elimination without partial pivoting can be written in Matlab [4] as:

```
function [L,U] = lu(A);
[m,n] = size(A);
L = zeros(n,n);
U = zeros(n,n);
for j = 1:n
    t = A(:,j);
    for k = 1:j-1
        if t(k) ~= 0
            t = t - t(k) * L(:,k);
        end;
    end;
    U(1:j,j) = t(1:j) ;
    L(j+1:n,j) = t(j+1:n)/t(j);
end;
```

This code will be our standard of elegance. Our goal is to let the programmer describe how to add partial pivoting and describe how to implement the algorithm efficiently, while keeping as much of the elegance of the above code as possible.

An optimized version of GEPP, written to a custom designed C++ sparse matrix library, is too long to include here, but the inner loop gives the feel of the whole:

```
for (SpaNzInorderIterator<Permutation,Range>
        iter(t, p, r1); iter; ++iter) {
  int k = iter.get_row();
  SparseVector L_col_k(L, k);
  spaxpy(t, p, NoRange, iter, L_col_k, p, NoRange,
        -(iter.get_value()), NoTrans);
}
```

This algorithm is as efficient as we could make it while still using library routines rather than open coding the entire algorithm. The code is a tangled mess. It is difficult to see the base algorithm in among all the other issues.

Among the additional issues addressed by this code are: The use of partial pivoting (all the 'p' arguments in the code), the merging of several logically distinct operations into a single spaxpy call, the use of special data structures, such as sparse accumulators [5]. Our goal is to address these issues in a way that avoids the tangling above.

An AOP solution

In AOP, a system is broken down into *components* and *aspects*. Components correspond to units of functionality; aspects are issues that cross-cut component boundaries. Stated another way, the component program describes the functionality of the system, that part normally thought of as the functional intent of the programmer, as separated from the less algorithmic concerns. These latter concerns are contained in aspect programs.

A complete AML program consists of the basic algorithm written in the component language, and annotations describing an efficient implementation of the component program. These annotations are written in several different aspect languages. Finally, there is a tool called an Aspect Weaver that takes all this code and produces an executable program. The remainder of this section describes first the component language, then each aspect language in turn along with the issue it addresses from our LU example. The last section describes the aspect weaver and performance.

The component language

We chose a component language that approximates the familiar Matlab language [4] for expressing the basic algorithm. Starting with Matlab, we added an additional iteration construct, for nzs, to explicitly express iteration through the non-zeros of a vector. We also imposed some restrictions to simplify compilation, such as not allowing different types of values to be assigned to the same variable.

Data representation aspect

The data representation aspect allows declarations along 5 different axes of data representation decisions:

Axis	Range
Element Type	integer, real, complex
Dimension	scalar, vector, row-vector, matrix
Representation	full, sparse, spa
Ordering	ordered, unordered
Orientation	by-rows, by-columns

The first two entries, element type and dimension, are the components of what is traditionally thought of as the type of the object. They describe what kind of data is visible to the base program.

The three other entries are additional representation information that is not visible to the base program. These other components determine the data structure that the implementation will use. They do not affect the behavior as seen from the base program, but they can dramatically affect performance. Representation specifies either full, sparse, or sparse accumulator (SPA). Ordering applies only to sparse or SPA representations, and indicates whether the non-zero elements should be maintained in order. Orientation applies only to sparse matrices, and indicates whether they should be stored by rows or columns.

Implicit permutation aspect

We found a need for an additional aspect to handle the implicit permutation of vectors and matrix rows and columns, which occur frequently in numerical code, to address issues like: ordering for sparsity [6], parallel partitioning [7], block triangular form [8], and effective preconditioning [9]. An implicit permutation declaration, written with a `view` annotation, like

```
view A through (p,:)
```

allows implicit permutations to be dealt with in one place, rather than being smeared throughout the code. It says that all references to the specified arrays should act as if the arrays were permuted as specified. Put another way, all indexing first goes through the permutation vectors. Changing the contents of the permutation vectors changes the effective permutation of the array.

Operator fusion aspect

The operator fusion aspect is responsible for indicating patterns of operations that can be efficiently performed as a unit by library routines. Unlike the other aspects, code for this aspect describes how to utilize the library in general, rather than how to implement a part of a particular application program. Since it is the library writer rather than the application programmer who writes the code for this aspect, we don't show it in this paper.

AML code for LU

The complete code for LU, using these languages, is:

```
function [L,U,p] = lu(A);
declare real sparse matrix A, L, U;
declare real          scalar v;
declare int           scalar m, n, j;
[m,n] = size(A);
L = zeros(n,n);
U = zeros(n,n);
declare permutation p;
p=[1:n];
view A,L,U through (p,:)
    for j = 1:n
        declare SPA t;
        view t through p
            t = A(:,j);
            for nzs k in order in t(1:j-1)
                t = t - t(k)*L(:,k);
            end;
            [v,piv] = max(abs(t(j:n)));
            piv = piv+j-1;
            p([j,piv]) = p([piv,j]);
            U(1:j,j) = t(1:j);
            L(j+1:n,j) = t(j+1:n)/t(j);
        end view;
    end;
end view;
```

The data representation declarations have been marked with a single bar to the left. Double bars mark code that deals with implicit permutations, including code in the component language that adjusts the permutation.

To our eye, this code does a good job of preserving the structure of the basic algorithm while expressing the desired efficient implementation.

The weaver

In most respects the weaver is a very simple compiler. Rather than trying to make smart inferences, the power of our weaver comes from the crucial domain information given by the most knowledgeable person: the user.

The weaver is built of several passes that can be seen as successive transformations applied to an Abstract Syntax Tree (AST) produced by the parser. Each pass consists of rewriting rules which are supported by a walker and a pattern matcher. These rules translate the AST into a new language. The language at each pass is nearly a proper superset of the language at the next pass (and thus of all later passes also). Therefore a particular pass can choose to rewrite part of the tree, or leave it for a later pass to deal with.

The first passes deal with information dissemination. The parser gives us a single AST that contains the component program, the implicit permutation aspect program, and the data representation aspect program. The weaver then propagates the permutation and representation aspects throughout the tree, according to their respective scopes. Since permutations stay in force across function calls, the weaver must note any component functions that are called with permuted arguments so that they can be automatically instantiated for the kinds of arguments passed at each point in the tree.

The latter passes of the weaving process can be roughly described as code generation. First there is a canonicalization pass, where the large number of possible AST shapes, many of which represent the exact same computation, are reduced to a smaller set of regularized forms. The weaver then applies the operator fusion aspect program. The operation fusion code consists of a set of translation rules, ordered by decreasing level of fusion. In a single top-down walk of the tree, each rule in turn is applied to the subtree at that point. If the rule matches, it completely translates that subtree into the output language, possibly recursing if it contains unfusable children. In addition to translating fused operations, simple operations are also translated by the same process. The whole code generator is run over the tree until the tree stops changing. At this point the whole tree has been translated into the final language, which is a transparent representation of C/C++. A post processor handles the conversion to C++, low-level compiling, and linking.

Results

The goal of AML is to allow sparse matrix code that is both easy to understand and modify, and efficient. This section measures the expression of LU in AML against those goals. We compare sparse LU written in AML with two other implementations: GPLU [10], a well-known Fortran version of the algorithm, also implemented (in C) in the Matlab version 4 internal library [5]; and SuperLU [11], the best performing research code, which uses much more complex algorithms to optimize cache behavior and to exploit structural symmetry.

Complexity

Quantifying the comprehensibility and modifiability of code is a notoriously difficult problem. There are many measures of code complexity [12], none of them entirely satisfactory. Using just the very simple measure of source lines of code, AML takes about 30 lines of code, GPLU takes 300 lines, and SuperLU takes 3000 lines. In other words, the AML code is an order of magnitude shorter than the GPLU implementation. The size comparison for the SuperLU code is not as pertinent, since it is performing a much more complicated algorithm.

Efficiency

Since the time to execute a sparse matrix code can depend on details of the actual matrix being manipulated, we measured the efficiency of the three implementations against a suite of standard test matrices from the Harwell-Boeing sparse matrix library [13]. The measurements were run on a SPARC 20, using the GNU compilers, running under SunOS 4.1.3.

The results, summarized in the table below, show that the AML implementation ranges from slightly slower than GPLU for easier matrices to essentially the same speed for hard matrices. Both are about 3 times slower than SuperLU. We include the comparison to SuperLU for completeness, even though it is not the kind of algorithm that the users of AML are likely to write. SuperLU deals with a complicated low-level issue that neither AML nor GPLU does, namely blocking data to exploit the memory hierarchy. One area of future work in AML would be to design an aspect to address this issue, hopefully yielding simpler code while retaining the performance of SuperLU.

#	name	size	nonzeros	AML (sec)	GPLU (sec)	SuperLU (sec)
1	gemat11	4929	33185	0.7	0.5	0.4
2	memplus	17758	99147	1.2	0.7	1.0
3	mcfe	765	24382	1.4	1.0	0.4
4	rdist1	4134	9408	4.1	3.3	2.1
5	orani678	2529	90158	4.2	3.8	0.6
6	jpwh991	991	6027	3.7	3.9	1.5
7	sherman5	3312	20793	6.2	6.0	2.1
8	lnsp3837	3937	25407	8.6	7.9	3.4
9	lns3937	3937	25407	9.3	9.0	4.0
10	orsreg1	2205	14133	13.2	12.9	4.3
11	sherman3	5005	20033	12.9	13.9	4.5
12	saylr4	3564	22316	23.1	24.8	7.5
13	goodwin	7320	324772	150.2	161.6	41.1

References

1. Kiczales, G., *et al.*, *Aspect Oriented Programming*. 1996, Xerox PARC: http://www.parc.xerox.com/spl/projects/aop/position.htm.

2. Kiczales, G., *et al. Aspect-Oriented Programming*. in *European Conference on Object-Oriented Programming*. 1997. Finland: Springer-Verlag.

3. Saad, Y., *Iterative Methods for Sparse Linear Systems*. 1996, Boston: PWS Publishing Company.

4. Mathworks, *MATLAB User's Guide*. 1992, The Mathworks Inc.

5. Gilbert, J.R., C. Moler, and R. Schreiber, *Sparse Matrices in MATLAB: Design and Implementation*. SLAM J. Matrix Anal. Appl., 1992. **13**: p. 333-356.

6. George, A. and J.W. Liu, *Computer Solution of Large Sparse Positive Definite Systems*. 1981: Prentice-Hall.

7. Shewchuk, J.R. and D.R. O'Hallaron, *Archimedes*. 1996: http://www.cs.cmu.edu/~quake/archimedes.html.

8. Pothen, A. and C.-J. Fan, *Computing the block triangular form of a sparse matrix*. toms, 1990. **16**: p. 303--324.

9. Duff, I.S. and G. Meurant, *The effect of ordering on preconditioned conjugate gradients*. BIT, 1989. **29**: p. 685--657.

10. Gilbert, J.R. and T. Peierls, *Sparse Partial Pivoting in Time Proportional to Arithmetic Operations*. SIAM J. Sci. Statist. Comput., 1988. **9**: p. 862-874.

11. Demmel, J.W., *et al. A Supernodal Approach to Sparse Partial Pivoting*. in *ILAY Workshop on Direct Methods*. 1995. Toulouse, France.

12. Henry, S. and D. Kafura, *Software Structure Metrics Based on Information Flow*. IEEE Transactions on Software Engineering, 1981. **SE-7**: p. 509--518.

13. Duff, I.S., R.D. Grimes, and J.G. Lewis, *Sparse Matrix Test Problems*. ACM Transactions on Mathematical Software, 1989. **15**: p. 1-14.

Client/Server Architecture in the ADAMS Parallel Object-Oriented Database System

Russell F. Haddleton and John L. Pfaltz

University of Virginia

Abstract. This paper describes issues encountered in the design and implementation of a parallel object-oriented database system. In particular, we find that the design of a client/server interface (that is, whether to use a page server or query server architecture) depends greatly on the expected application environment. We believe that the query server model is more appropriate for most scientific database applications. This paper develops the reasons for this assertion. Then we discuss the implementation of a working parallel object-oriented, database system, called ADAMS, that has been developed at the University of Virginia. ADAMS is best known as the database system underlying the "Oracle of Bacon" website, which is currently responding to over 5 million queries per year. We show that set operations such as union and intersection can be performed in a completely data-parallel fashion, and that implicit join queries can be performed with very few inter-processor messages in a way that scales well. All data access in the system can be completely monitored, so these observations are supported by experimental results. Finally, we compare our approach with an equivalent page server version.

1 Introduction

The design of the client/server interface in object-oriented database systems has received considerable recent attention [1,4]. In creating ADAMS, a parallel object-oriented database system, the choice of an appropriate client/server architecture was a major design decision. This design has been affected by hardware considerations, application considerations, and the expected level of user activity. Considerable effort was devoted to implementing an effective client/server architecture.

ADAMS uses a query server architecture. In this paper, we explain the reasoning behind this design decision. Analytic modeling and actual performance results provide further support for our choice. But we do not argue that this architecture is best in all situations. The factors we considered for anticipated ADAMS applications drove us to our particular design.

2 Client/Server Architectures

By a page server model we mean a configuration in which one or more server processors directly access secondary storage at the behest of other processes which are their clients. Based on its own internal logic, a client process requests a page denoted by some persistent storage address. The server locates the page in storage and transmits it to the client, which then extracts the relevant portions. A page server has no comprehension of the contents of a page, whether it consists of objects, tuples, or perhaps code fragments. Page server architectures have been a frequent choice in object-oriented database implementations [4]. The server code is very simple, and features such as page level locking and transaction rollback add little complexity. According to Franklin [4]: "data-shipping approaches offload functionality from the server to the clients. This may prove crucial for performance, as the majority of processing power and memory in a workstation-server environment is likely to be at the clients." (Data-shipping is another name for page server.)

The object server model is similar, except that the client requests individual objects which the server extracts from the appropriate pages and returns. The server must know about object placement in pages. According to DeWitt, *et al.*[1], in an object server architecture object methods can be applied at either the client or server, allowing for some variation between implementations. Besides reducing the amount of data that must be shipped between the server and its clients, recovery and concurrency control can be handled at a finer granularity.

The query server model adds further server complexity. A client process sends operations, commands, or code fragments to the server along with any necessary operands. The server executes the operations and may, or may not, return the results of the operation to the client. The server not only uses object-level logic, it can exploit indices and may make optimization decisions.

An OODB implementation need not fit precisely into one of the above categories. As Wilcox [12] says, "the variety displayed in OODB implementations is at once confusing and encouraging". Yet whatever the label, some complexity is provided by the processor directly used by the client, and some by one or more server processors. The above labels capture this range of choices well.

How does one choose among these architectures? We contend that the choice should be governed by the nature of the intended applications. We will argue further that for many scientific database applications the query server architecture is optimal.

3 Scientific Database Applications

There are at least three fundamental differences between the way scientific data and business data are processed. First, commercial processing is characterized by many transactions originated by many clients that examine or update the current state of the database. In contrast, a scientific measurement, once inserted into the database, is rarely changed. We do not see the posting of small transactions by many users as the major scientific activity. Rather we expect a handful of researchers to execute fairly complex queries in search of phenomena of interest.

Second, a typical commercial transaction deals with only a few data items that are well identified by an account number, a social security number, or the like. Relatively few elements of scientific data are uniquely identified in this manner. Sometimes a scientist will be interested in a particular pixel in a particular image, but frequently he will want all the pixels of a specified region to serve as input to some computation, possibly statistical. Other common queries are "volumetric;" that is, they denote all elements satisfying some specified range constraints. The following query expression is a good example.

$$S \leftarrow \{x|\; p1 \; < \; x.\text{pressure} \; < \; p2 \; \text{OR}$$
$$t1 \; < \; x.\text{temperature} \; < \; t2 \; \text{AND}$$
$$x.\text{instrument.elevation} \; > \; 1000\}$$

Here, we are looking for all objects, x, whose observed pressure falls between the given limits, or whose temperature lies in its specified interval and the observations were recorded on a sensing instrument sited above 1000 meters. We will consider this example more fully in Sec. 4.3.

This query shows the third fundamental difference; it may access thousands of data items, whereas a single transaction will normally access fewer than ten.

These three differences are sufficient to suggest that query servers are more appropriate for scientific work. First, the convenience of page granularity for transaction atomicity and concurrency control is much less important. Second, complex query resolution, or traversal of multi-link access paths, is more easily accomplished at a server which is "close" to the data than by a client which is well "removed" from it. And finally, exchange of very large quantities of data, whether pages or objects, between the server and client can swamp any communications network. These three considerations were sufficient to incline us towards a query server architecture.

The value of the object model in scientific database work is a frequent theme [8]. But object-oriented databases have been seen as poor performers when handling large data bases [11]. It seems certain that high-performance, scientific database implementations will have to be highly parallel. Parallel object-oriented database systems introduce two new considerations. First, when a system is distributed over n processors, it becomes likely that the aggregate of server memory and CPU resources will exceed those of active clients. In a single-server system that accommodates multiple clients, a page server approach makes sense because the client, often a powerful workstation, can easily provide the cache storage to receive pages, and the cycles needed to transform these pages into objects, at a rate fast enough to satisfy the user's needs [4]. But, with multiple servers feeding only a few clients, even relatively small query results from a large database can swamp a client cache. Simple load balancing suggests that when the number of servers exceeds the number of clients, the processing load should be redistributed.

The second consequence of implementing a parallel system is that as the number, n, of processors increase, so too must the inter-processor message traffic. Because "shared-memory and shared-disk do not scale well on database applications"[2], our focus has been on shared-nothing implementations. The key to effective parallel code in shared-nothing implementations is to minimize inter-process message passing. Page servers which transfer entire pages when only a few objects will be used, and algorithms that rely on extensive inter-server data transfer, can cause severe bottlenecks.

Therefore we chose a query server architecture and an object distribution methodology which minimize inter-server communication.

4 ADAMS Architectural Overview

In this section we flesh out some of the more important details associated with the ADAMS implementation. ADAMS [10] was initially designed to support highly concurrent scientific computation [5] on shared-nothing, message-passing processors. The performance reported in this paper has been observed in a system running on a cluster of 8 SPARC2 processors, each with an attached Seagate 1.2 GB disk and Ethernet connection network. It is object-oriented, with user-defined data classes. Each instantiated object within one of these classes is assigned a persistent, unique object identifier, or *oid*. These *oids* serve as object surrogates throughout the system and play a prominent role in the way we distribute objects and minimize inter-processor communication.

4.1 ADAMS Client/Server Details

An ADAMS server is essentially a query server. It executes set and index level operations. But at times it also functions as a very fine-grained data or object server that can honor requests not for entire objects, but individual attribute values of objects. This handles an object server problem mentioned by DeWitt, *et al.*: "since the software on a workstation simply requests an object from the server without generally being aware of the size of the object, large, multi-page objects may move in their entirety, even if the application needed to access only a few bytes" [1]. Both client and server processes have been implemented in a layered fashion, as shown in Fig. 1. The user application program of the client process

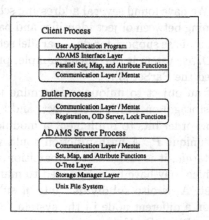

Fig. 1. ADAMS process architecture.

is written in some host language, such as C, C++, or Fortran, with embedded ADAMS statements. These ADAMS statements are translated by a preprocessor into appropriate calls to ADAMS run-time routines residing in the ADAMS interface layer. The database operands of these run-time procedures are always denoted by object *oid*. But neither the user, nor the ADAMS preprocessor, need be aware of the way sets of objects and their attributes may have been distributed. We have deliberately created a seamless, parallel implementation.

The set maintenance and attribute access level of the servers, shown in Fig. 1, constitutes the heart of the ADAMS system. ADAMS uses a "decomposed storage model," or DSM [7], in which the attribute values of an object are stored as (*oid, value*) pairs. All references to attribute values for a named attribute, for example pressure" above, are stored in a B-tree associated with that attribute. Thus there is a "pressure" B-tree, indexed by *oid*. All sets and all attributes are maintained as separate trees in order to facilitate access. The inverses of these attribute trees, indexed by value, are created as needed. Each server maintains its own set of trees. Below this level, we show a storage manager level which was implemented on top of the Unix file system. We describe how this model has been used in a parallel form in the next section. The use of DSM makes schema evolution fast and efficient, while making the storage and access of data associated with complex or large objects a simple task.

One secret for minimizing message traffic lies in the nature of the messages themselves. For example, with sequences of low granularity operations, such as set insertions or object attribute assignments, we are frequently able to bundle many operations for a particular server into a single message. As these messages do not require a response from the servers, the message burden is kept low when loading large quantities of data, such as that from a remote sensor stream.

4.2 Partitioning by *oid*

Processing very large sets of scientific data requires the distribution of these sets over multiple disks and processors. It seems to us that a horizontal partitioning of sets by *oid* is natural. We have found several addressing schemes in the literature that use a simple mapping between object identifier and partition, although only a few claims of direct database support for data parallel set operations have been encountered. The SHORE system is one such example, providing support for a parallel structure called the ParSet [3].

We use the *oid* of an object to uniquely determine the partition P_j (that is, the processor and storage) on which the object, and all its attribute values, resides. We use the low-order bits to implement a modulo operator which partitions every *oid* to a unique P_j. A hash function could work, as well. A set is an object. Based on its *oid*, it is assigned to a partition, P_j. But the members of a set are objects which may have been assigned to many different partitions. Consequently, each set, A, is also subdivided into n subsets, $A_1,..,A_n$, where each subset, residing on a different node in the system, is comprised of just the objects assigned to partition, P_j. If a particular object x is a member of two ADAMS sets, say $x \in A$ and $x \in B$, x will always be represented as a member in two subsets on the same partition subset. That is, if $x \in A_j, 1 \le j \le n$, and $x \in B$, then $x \in B_j$ as depicted in Fig. 2. Using such a scheme, large granularity

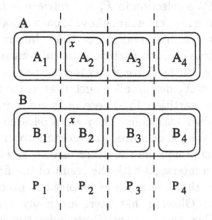

Fig. 2. Distribution of sets A and B.

set operations, such as union, intersection, and difference, can be performed in a data parallel manner, as can selection by attribute value, as in p1 < x.pressure < p2. ADAMS makes abundant use of these data parallel set operators.

An essential characteristic of the object-oriented model is that attributes may be object-valued. The term x.instrument.elevation < 1000 is representative. The attribute **instrument** denotes a sensor object y with elevation attribute greater than 1000 meters on which the observation x was made. Queries involving object-valued attributes are somewhat more complex to process because the object y may very well have been assigned to a different partition than x. The presence of object-valued attributes, or implicit joins, forces inter-server communication. It is essential to reduce their impact on performance.

4.3 Query Processing Details

We consider queries such as the one given in Sec. 3. The logical connectives AND and OR can be naturally implemented as set operators *intersect* and *union*. Consequently the approach we take is to find all x satisfying the first term, p1 < x.pressure < p2, and all x satisfying the second term t1 < x.temperature < t2. These two sets are unioned. This will then be intersected with the last term.

Parallel selection with respect to an attribute is well understood; each processor searches the indexes for the objects on its partition that satisfy the criteria. It is a completely data-parallel operation, with no cross talk between the processors. The conjunct which involves an object valued attribute is a bit different.

Like all attributes, an object-valued attribute is a single-valued function. But when applied to an object, the attribute function returns a reference to another object instead of a value. Also like any other attribute, the system indexes these object-valued attributes using the referenced *oids* as the key values and the referencing *oids* as the entries. When partitioning data by *oid*, we partition the attribute index trees by the referenced *oids*. That means that if object oid_k on P_k references oid_j on P_j, a selection in P_j will retrieve oid_k.

To process the term x.instrument.elevation > 1000, the system first selects all sensor objects y with elevation > 1000. In our database there were approximately 60 of these on each partition. In an iterative loop over each object, actually each *oid* y, all the measurements x are retrieved that have that *oid* as instrument value. Let X_j denote all x such that x.instrument = y_j, where y_j is a sensor object on partition P_j. These *oid*'s comprising X_j must now be distributed to their proper partitions. They are separated into batches, e.g. all oid_k bound for P_k, and then transmitted[1] to their appropriate processors. Server P_j in turn receives its subsets from the other servers, assembles them all into a single set, which it then intersects with the result of the first two terms.

Table 1 summarizes the data accesses involved in executing the query in a million-object database. Observe that there are many more random, than sequential, disk reads. When an inverse attribute index tree is created, it is almost

Table 1. Data accesses in the query statement.

	P_1	P_2	P_3	P_4	Sum
| p1 < pressure < p2 |	7,466	7,499	7,503	7,439	29,907
| t1 < temperature < t2 |	47,631	47,657	47,371	47,591	190,250
| y.elevation > 1000 |	61	71	70	49	251
| result |	19,179	19,079	19,120	19,052	76,430
Random disk reads	1,872	1,868	1,800	1,673	7,213
Random disk writes	25	26	25	26	102
Sequential disk reads	643	373	268	265	1,549
Sequential disk writes	1,173	1,167	1,161	1,173	4,674
Data messages sent	54	65	63	45	227
Data messages received	57	54	55	61	227

impossible to cluster the nodes, so virtually all disk accesses requires random head movement. In this query, nearly all of the sequential disk operations arise from the sorting of *oids*. To facilitate the set operations, all sets are kept in lexicographic *oid* order. But, the *oids* returned from selection over an attribute are not in order and must be sorted. With a bit of cache management, described in [6], we can sort surprisingly large sets completely in cache; but for safety, we always sequentially flush the entire cache to disk on completion. Only occasionally must portions of these sets be sequentially re-read from disk.

Note that although 76,430 observation objects were eventually retrieved from 1,000,000 observations, most of the processing was data-parallel. At most 65 inter-server data messages were sent by any server. In addition, the client sent control messages initiating the query, and there were a handful of short control

---

[1] In ADAMS, each data message can contain up to 1,000 *oids*.

messages between servers that coordinated the data exchange. We have been quite successful in minimizing inter-processor message traffic.

5 Performance

Because set operators are completely data parallel, they exhibit linear speed up, which is defined t_n(task of size k) = t_1(task of size k)/n where t_n denotes execution time in an n processor system. The complex query, on the other hand, exhibits *super*-linear speed up, as shown in Fig. 3, where theoretical linear speed up is shown by the dashed line.

It is not hard to show that, asymptotically, such a query cannot exhibit linear speed up because (1) its inter-server message traffic must increase as the number, n, of processors increases; (2) the necessary sort operations are O(n lg n); and (3) the analytic cost expressions contain constants that do not scale at all [6]. Nevertheless, we get apparent super-linear behavior because both searching inverse attribute trees and sorting are cache dependent. When we double the number of processors, we are more importantly doubling the available cache. For these reasons, scale up is regarded as a much more accurate measure of parallel performance, even if it is harder to derive.

A system exhibits linear scale up if t_n(task of size n*k) = t_1(task of size k), that is, if the number of processors is doubled the size of the task is doubled. The problem is that to fairly double the size of a task, one must really double the size of the database in which it is embedded as well. This is possible only with experimental databases such as ours. We created databases of 125000, 250000, 500000, and 1000000 observation objects respectively, and ran the same queries. Each query, or operator, was repeated 10 times, always using a "cold" system. Figure 4 shows the scale up performance of the query. It is not quite linear, although the rise is almost imperceptible. We believe this kind of performance is due to our success in minimizing inter-processor message traffic, both between client and servers and between the servers themselves.

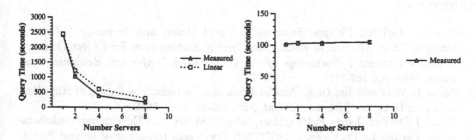

Fig. 3. "Super-linear" query speed up.   **Fig. 4.** Scale up of the query.

6 Conclusions

In a scientific environment in which complex volumetric queries are the norm, a query server architecture seems desirable. We have shown that this architecture scales well in our parallel implementation.

We wish to make a comparison with an equivalent page server architecture. This is a bit difficult since, in a real page server system, a different algorithmic approach may be adopted. Nevertheless some rough comparisons can be made. The objects in our test database were approximately 100 bytes long. So, given our 8K pages, we can expect about 81 objects per page, or approximately 12,345 pages in all. The retrieved set of 76,430 objects in our running query comprises about 7.6% of the entire million-object database. There is no reason to expect any significant clustering of these retrieved objects. So if we assume a reasonably uniform distribution of these objects, there will be about 6.1 desired objects per page. Just to return the results of this query in the form of pages will require transmission of nearly all 12,000 pages. This is why in Sec. 3, we showed that the sheer size of query results in scientific database applications constituted a reason for considering a different client/server architecture.

Orenstein, *et al.* describe query processing in a page server system. They show that "... queries execute on the client. This does not mean that the entire contents of a collection, or index, is sent to the client in order to evaluate a query. As with other objects, only those pages containing referenced addresses are fetched from the server" [9]. Just transmitting the relevant index pages to the client in a sequential fashion so it can conduct the search is expensive. Our running query required search on three attributes. This is why, in Sec. 3, we asserted that selections should be conducted in the server "close" to the data.

We have shown how, in the case of the ADAMS object-oriented database system, the choice of scientific database as an application area has resulted in a number of design and implementation decisions reflected throughout the system. It is possible that focusing on other application areas would have yielded different design choices. Yet the current design has been effective, as supported by a wide range of performance tests.

References

1. David J. DeWitt, Philippe Futtersack, David Maier, and Fernando Velez, "A Study of Three Alternative Workstation-Server Architectures for Object Oriented Database Systems", *Proceedings of the 16th VLDB Conference*, Brisbane, Australia, 1990, pp. 107-121.
2. David DeWitt and Jim Gray. Parallel Database Systems: The Future of High Performance Database Systems, *CACM*, 35(6):85-98, June 1992.
3. David J. DeWitt, Jeffrey F. Naughton, John C. Shafer, and Shivakumar Venkataraman, "ParSets for Parallelizing OODBMS Traversals: Implementation and Performance", *Proceedings of the Third International Conference on Parallel and Distributed Information Systems*, Austin, TX 1994 Texas,1994.
4. Michael J. Franklin, *Client Data Caching: A Foundation for High Performance Object Data Systems*, Kluwer Academic Publishers. Boston, MA. 1996

5. Andrew S. Grimshaw, John L. Pfaltz, James C. French, Sang H. Son, Exploiting Coarse Grained Parallelism in Database Applications, *PARBASE-90 Intern'l Conf. on Databases, Parallel Architectures and their Applications*, Miami Beach, FL, 1990, pp. 510-512.
6. Russell F. Haddleton, Parallel Set Operations in Complex Object-Oriented Queries, Ph.D. Dissertation, Univ. of Virginia, Aug. 1997.
7. Setrag Khoshafian, *Object-Oriented Databases*, John Wiley and Sons, New York, 1993.
8. David Maier and David M. Hansen, "Bambi Meets Godzilla: Object Databases for Scientific Computing", Proceedings of the Seventh International Working Conference on Scientific and Statistical Database Management, Charlottesville, VA 1994, pp. 176-184.
9. Jack Orenstein, Sam Haradhvala, Benson Margulies, Don Sakahara. Query Processing in the ObjectStore Database System, Proc. 1992 ACM SIGMOD Conf., San Diego, CA, June 1992.
10. John Pfaltz and James French, Scientific Database Management with ADAMS, Data Engineering, 16, 1 March 1993, pp. 14-18.
11. Bindu R. Rao, *Object-Oriented Databases: Technology, Applications, and Products*, McGraw-Hill, New York, 1994.
12. Jonathan Wilcox, Object Databases: Object methods in distributed computing, Dr. Dobb's Journal, 222, November, 1994. pp. 26-32.

Pattern–Based Object–Oriented Parallel Programming

Steve MacDonald, Jonathan Schaeffer, and Duane Szafron

University of Alberta, Edmonton, Alberta CANADA T6G 2H1
Email: {stevem,jonathan,duane}@cs.ualberta.ca

1 Introduction and Motivation

Over the past five years there have been several attempts to produce *template-based* (or, as they are now called, *pattern–based* [4]) parallel programming systems (PPS). By observing the progression of these systems, we can clearly see the evolution of pattern–based computing technology. Our first attempt, Frame-Works [9], allowed users to graphically specify the parallel structure of a procedural program in much the same way they would solve a puzzle, by piecing together different components. The programming model of FrameWorks was low–level and placed the burden of correctness on the user. This research led to Enterprise [7], a PPS that provided a limited number of templates that could be composed in a structured way. The programming model in Enterprise was at a much higher level, with many of the low–level details handled by a combination of compiler and run–time technology.

In this progression, we can see more emphasis placed on the *usability* of the tools rather than raw performance gains [12]. Each successive system reduced the probability of introducing programmer errors. However, since performance considerations cannot be ignored, each successive systems also supported incremental application tuning. A related performance issue is *openness* [10], where a user is able to access low–level features in the PPS and use them as necessary. This work led to a critical evaluation of pattern–based systems that provides the motivation for our new system, CO_2P_3S (Correct Object–Oriented Pattern-based Parallel Programming System, pronounced "cops").

In this paper, we present an architecture and model for CO_2P_3S in which we address some of the shortcomings of FrameWorks and Enterprise. Our continuing goal is to produce usable parallel programming tools. The most important shortcoming we address is the loose relationship between the user's code and the graphical specification of the program structure. Enterprise improved on Frame-Works by verifying a correspondence between the parallel structure and the code at compile–time. However, we feel that forcing the user to write a program that conforms to an existing diagram is redundant. If the structure of the application is already known, then the basic framework can be generated automatically. This reduces the amount of effort required to write programs, while simultaneously reducing programmer errors even further. The CO_2P_3S architecture also supports improved incremental tuning. The architecture is novel in that it provides several user–accessible layers of abstraction. At any given time during performance tuning, a programmer can work at the appropriate level of abstraction, based on what is being tuned. This can range from modifying the basic parallel pattern at the highest level, to modifying synchronization techniques at the middle layer, to modifying which communication primitives are used at the lowest level. Our goal is an open system where the performance of an application is directly commensurate with programmer effort.

2 The Architecture of CO_2P_3S

The architecture of CO_2P_3S consists of three layers: **Patterns**, **Intermediate Code**, and **Native Code**. These layers represent different levels of abstraction, where the abstraction of each layer is implemented by the one underneath.

Each layer is transformed during compilation to the layer underneath. At the pattern layer, the developer selects a pattern using a graphical tool. The PPS then generates a template for the parallel application and the user is restricted to providing sequential application–specific code at particular locations in the generated template. At the pattern level, the PPS guarantees the correctness of the generated program by using conservative synchronization mechanisms that may not yield peak parallel performance.

Unlike Enterprise, CO_2P_3S provides programmer access to the second layer where the templates can be edited. From this layer down, the programmer is responsible for the correctness of the resulting program. To simplify this task, we provide a high–level parallel programming language that is an extension of existing OO languages (Java or C++). This superset includes keywords to denote parallel classes, specify concurrent activities and express necessary synchronization (using constructs such as asynchronous methods, threads, and futures).

Finally, the third layer is the native programming language augmented with a library that provides the services required by the first two layers. Users are given full access to all language features and library code.

We believe that providing intermediate levels of abstraction can provide several benefits. First, by generating correct template code at the top level, we can ensure that a user has a working parallel program before the tuning process begins. Second, it eases the tuning process by introducing the run–time system in smaller increments. These smaller increments provide better opportunities for novice or intermediate users to find a comfortable level of abstraction while still providing full access to the run–time system for experienced users. Lastly, it should always be possible to improve the performance of a program by using the abstraction of a lower level. If so, then the performance of an application should be more directly commensurate with programmer effort.

3 The Model

In this section, we more fully specify the model and demonstrate it using an example program. Our example shows some details of a generic mesh computation.

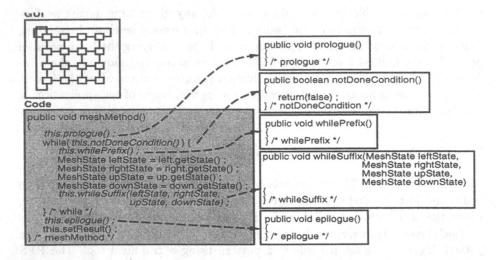

Fig. 1. The first layer of CO_2P_3S. Shaded code is generated by the system and cannot be modified. Italicized methods are null and must be implemented by the user. Skeleton implementations are shown on the right.

The first layer specification is shown in Figure 1. This layer consists of both a graphical representation of the pattern and the accompanying generated code. The pattern itself has several parameters that must be given. The parameters include the width and height of the mesh and its boundary conditions. The boundary conditions are the most interesting of these parameters since they can affect the user code. The reason for this can be seen in the accompanying code fragment. It shows the main loop in the evaluation of the mesh, with the shaded part representing the generated code and italicized code representing hook methods that the user must implement. In the hooks, the code uses all four mesh neighbours. However, if the mesh is not fully toroidal, then not all the neighbours will exist and the supplied code is not applicable. To correct this, we allow the user to specify different code for the different special cases. In this paper, for simplicity, we will assume the mesh is fully toroidal.

One critical feature of the first layer is that the generated code encapsulates the communication and only allows the user to operate on the data. This feature allows our PPS to make certain correctness guarantees, such as ensuring that each element of the mesh is referenced. The code in Figure 1 demonstrates this property by getting the state from the mesh neighbours and communicating the results of the computation to a *collector* object responsible for collecting the results (encapsulated within `setResult()`). Another critical aspect of this code is that we provide sufficient hooks for users to completely implement their programs. This requirement must be evaluated on a pattern–by–pattern basis. For the portion of the mesh in Figure 1, we provide five hooks. `prologue()` and `epilogue()` are executed before and after the mesh computation and can be used for initialization and cleanup, such as implementing instrumentation.

`whilePrefix()` and `whileSuffix()` are executed for every iteration of the mesh and can be used to preprocess local data and perform the desired mesh operation. Finally, `notDoneCondition()` specifies the terminating condition for an element of the mesh. The mesh stops executing when all the elements are finished. The default implementation of these methods is given on the right side of Figure 1.

From the pattern layer code, we generate the intermediate code of the second layer shown in Figure 2. The figure also contains an example of a method operating on data in the mesh, which would have been added by the user at the first level. For the mesh, the keywords `left`, `right`, `up`, `down` are defined to refer to the neighbours. The keyword `barrier` is also required to define the necessary synchronization for the mesh and is automatically inserted in the transformation of the mesh code from the first to the second layer. Other, more general keywords are also supported by CO_2P_3S.

```
public void meshMethod()
{
    this.prologue() ;
    while (this.notDoneCondition()) {
        this.whilePrefix() ;
        MeshState leftState = left.getState() ;  MeshState rightState = right.getState() ;
        MeshState upState = up.getState() ;  MeshState downState = down.getState() ;
        barrier ;
        this.whileSuffix(leftState, rightState,  upState, downState) ;
    } /* while */
    this.epilogue() ;
    this.setResult() ;
} /* meshMethod */

public void whileSuffix(MeshState leftState, MeshState rightState,
                        MeshState upState, MeshState downState)
{
    // Take the average of four neighbours and gather timing information.
    this.value = (this.value + leftState.getValue() + rightState.getValue() +
                  upState.getValue() + downState.getValue()) / 5.0 ;
    this.stopTimer() ;
    this.gatherStatistics() ;
} /* whileSuffix */
```

Fig. 2. Some code from the second layer of CO_2P_3S. The italicized code are keywords inserted by the transformation between layers.

A feature of the generated code is that it is fully functional even if the user does not implement any of the hooks. In the mesh example, the default implementation for `notDoneCondition()` returns false so the mesh immediately exits. The main body of the mesh, if invoked, will retrieve the state from each of its neighbours and perform a null operation. By producing code that is capable of executing, we can verify that our patterns are correct and that the user will start with a parallel program that does not contain any communication errors.

The user can also edit the generated code at this level, which allows the structure of the mesh to be extended or modified. For example, we can add new neighbours that were not defined in the original mesh pattern. We can use this feature to define an arbitrary stencil to be convolved over the mesh.

Finally, from the second layer code of Figure 2, we generate the native code given in Figure 3. Native code replaces the keywords inserted in the second layer with library calls implemented by our run–time library. This replacement may simply replace the keyword with an accessor, such as accessing the neighbours of the mesh. Others may be more complex, such as replacing **barrier** with a call to the thread group for the threads executing the mesh. Users can also use any available library call or language feature.

```
public void meshMethod()
{
    this.prologue() ;
    while (this.notDoneCondition()) {
        this.whilePrefix() ;
        MeshState leftState = this.getLeft().getState() ; MeshState rightState = this.getRight().getState() ;
        MeshState upState =this.getUp().getState() ; MeshState downState =this.getDown().getState() ;
        this.getMeshThreadGroup().barrier() ;
        this.whileSuffix(leftState, rightState, upState, downState) ;
    } /* while */
    this.epilogue() ;
    this.setResult() ;
} /* meshMethod */
```

Fig. 3. The third layer of CO_2P_3S. The italicized code represents replaced keywords. The code for other methods is accessible but not shown.

4 Evaluating the Model

In this section, we examine the 13 desirable characteristics of pattern–based PPSs described by Singh *et al.* and apply them to CO_2P_3S [10]. Like Singh *et al.*, we break the characteristics into three categories. We also use their short names which are shown in bold in this text.

Structuring the Parallelism: This category examines how users can structure the parallelism in their programs. There should be as few restrictions as possible.

CO_2P_3S addresses **separation** by expressing the parallelism diagrammatically and allowing application–specific code to be inserted into the pattern. In our case, the diagram is a *collaboration diagram*. Further, the CO_2P_3S approach of generating code that invokes user hooks provides more opportunity to re–use the hooks when the user changes the pattern. Since the hook code does not affect the communication flow of the program (because the communication code is generated and cannot be edited by the user), there is a greater possibility that this code can be used in another pattern.

We can allow the parallelism in a program to be **hierarchically** specified by allowing patterns to be substituted for the sequential components in a program. This composition can be compared to the Composite design pattern [4]. It provides a structured way of building complex program elements.

We can only address **independence** and **utility** once the set of templates has been decided upon. Typically, independence has been addressed by rigourously defining the inputs and outputs of each design pattern, or by creating separate processes so that each pattern has only one input and one output. Either strategy is applicable here.

Currently, this research has not focused on how to provide a way of **extending** the set of existing patterns. There has been other work in this area, such as DPnDP [11]. We hope to use and continue the work started by this system.

The major contribution of our architecture is how it addresses the problems of **openness** and **correctness**. The pattern layer addresses correctness by generating template code from the design patterns. Since the generated code cannot be modified, we can strictly enforce the structure of the program. Combined with a run–time library, this code can offer a high–level model that frees the user from the low–level details of parallel programming and guarantees the correctness of the parallel structures.

The subsequent layers of the system are intended to address the problem of openness by providing successively more access to both the generated code and the run–time system. The intermediate layer provides more control over the user program by providing access to the generated code of the first layer, but still provides a high–level programming model for easier programming. The user can optimize the generated code, but is then responsible for its correctness. Finally, the last layer provides access to the complete programming system.

Programming: This category evaluates the style and structure of the application code written by the user.

In examining the language characteristics, we should emphasize that the CO_2P_3S architecture is intended to be independent of language. As such, it should be possible to use the architecture for a variety of languages, satisfying the goal of using an existing, **familiar programming language**. The definition of this characteristic also includes the ideal of preserving the semantics and syntax of this programming language. We purposefully stray from this ideal, though, as we feel that preserving the semantics of a sequential language unnecessarily limits the potential concurrency. As a small example, consider programming languages that support run–time exceptions. To fully preserve the sequential semantics, it is necessary to execute every statement in order. However, concurrent execution could execute code that would not be executed in the event of an exception [14]. Without concurrent execution, computational parallelism is useless. A limited set of new keywords can provide a usable abstraction for parallel programming without greatly disturbing the rest of the language. We feel that the benefits of a well–planned abstraction that is properly integrated into the language can outweigh the risks of modifying a programming language. Nevertheless, we are only proposing these modifications at the intermediate code layer of our PPS with the understanding that at the native code layer, they will be translated to a standard programming language; in this case, Java or C++.

Finally, we anticipate that CO_2P_3S will fail to meet the **non–intrusiveness** characteristic. As noted by Singh et al., the only way to fully address this problem is to implement a compiler that automatically generates a parallel program from sequential code (which also solves the language objective). Unfortunately, current compiler technology cannot create coarse–grained parallel programs.

User Satisfaction: This last category focuses on a combination of performance and usability concerns.

Without an implemented and mature system, most of the characteristics in this last category cannot be evaluated. In particular, we cannot address the **support** and **usability** objectives. The **performance** objective is dealt with by our architecture for incremental tuning. **Portability** can be addressed by providing different implementations of the Native Code layer for each architecture.

5 Related Work

In addition to FrameWorks and Enterprise, there are several other graphical parallel programming systems. Mentat [5] and HeNCE [2] both represent programs as directed graphs. The programming model of Mentat is similar to that of Enterprise with some extensions for C++. However, neither system is pattern–based and neither supports incremental tuning. The parallel programming language P^{3}L [1] provides a set of design patterns that are composed to create programs. The programming model involves explicit communication that is type–checked at compile–time. Unfortunately, new languages impose a steep learning curve on new users. Also, the language is not object–oriented and does not support multiple abstractions. The DPnDP system is similar to the Mentat system except that the nodes in the graph may be implemented using design patterns. This project also provides an interface for adding new patterns [11].

There has also been some work done in verifying code based on design patterns. Sefika *et al.* [8] proposed a model for verifying a program's adherence to a design pattern using a combination of static and dynamic program information, and also suggest generating code from a pattern. The generation of code from a design pattern has been done for the patterns in [4] by Budinsky *et al.* [3] This project delivers the source code that implements each design pattern. This code is then modified by the user for its intended application. In contrast, we do not allow the user to immediately edit the generated code, so we can more rigourously enforce the constraints of the selected pattern.

6 Preliminary Results and Current Status

We have some preliminary results based on a hand–coded implementation of a mesh similar to that in Section 3. The example program we use is a reaction–diffusion texture generator [13]. The algorithm can be described as two interacting LaPlace equations, which simulate the reaction and diffusion of two chemicals over a surface. The result is a texture map that approximates zebra stripes. We solve the problem using straightforward convolution.

The program was executed on a SUN Ultra–SPARC 2 with 2 200MHz Super-SPARC processors and 128M of main memory. The program was a hand written Java program that implements the mesh pattern described in this paper. We did not perform any optimizations on the communication structure itself, although we made some optimizations at the user code layer to reduce the amount of state moved between threads. The sequential program is a modified version of the mesh code tuned for the case where there is only one mesh element. The threaded version was run using native threads (multiple threads using both processors) and achieved a wall clock speed–up of 1.26. Although this speed–up is not very good, Java native thread interpreters are new technology and it should improve in the future. The synchronization requirements of the problem is also a limiting factor.

We are currently implementing CO_2P_3S in Java. However, we are being careful to keep the architecture and model independent of any specific language. We will support a rich set of patterns for the first layer of the system, including meshes, trees, pipelines, master/slave, and iterators. The target architecture is a multiprocessor machine with shared memory executing threads in parallel. This is currently supported for Solaris 2.5.1. We may also investigate using networks of workstations with a message–passing package such as Voyager [6].

References

1. B. Bacci et al. P^3L: A structured high level parallel language and its structured support. *Concurrency: Practice and Experience*, 7(3):225–255, May 1995.
2. A. Beguelin et al. HeNCE: A hetergeneous network computing environment. Technical Report UT-CS-93-205, University of Tennessee, August 1993.
3. F. J. Budinsky, M. A. Finnie, J. M. Vlissides, and P. S. Yu. Automatic code generation from design patterns. *IBM Systems Journal*, 35(2):151–171, 1996.
4. E. Gamma, R. Helm, R. Johnson, and J. Vlissides. *Design Patterns: Elements of Reusable Object–Oriented Software*. Addison–Wesley, Reading, Mass., 1994.
5. A. S. Grimshaw. Easy to use object–oriented parallel programming with Mentat. *IEEE Computer*, 26(5):39–51, May 1993.
6. ObjectSpace, Inc. *Voyager Version 1.0 Beta 2.1*. http://www.objectspace.com.
7. J. Schaeffer, D. Szafron, G. Lobe, and I. Parsons. The Enterprise model for developing distributed applications. *IEEE Parallel & Distributed Tech.*, 1(3):85–96, 1993.
8. M. Sefika, A. Sane, and R. H. Campbell. Monitoring compliance of a software system with its high–level design models. In *Proceedings of the 18th International Conference on Software Engineering (ICSE-18)*, pages 387–396, March 1996.
9. A. Singh, J. Schaeffer, and M. Green. A template–based approach to the generation of distributed applications using a network of workstations. *IEEE Transactions on Parallel and Distributed Systems*, 2(1):52–67, January 1991.
10. A. Singh, J. Schaeffer, and D. Szafron. Experience with template–based parallel programming. *Concurrency: Practice and Experience*, 1997. To appear.
11. S. Siu, M. De Simone, D. Goswami, and A. Singh. Design patterns for parallel programming. In *Proceedings of the 1996 International Conference on Parallel and Distributed Processing Techniques and Applications (PDPTA'96)*, August 1996.
12. D. Szafron and J. Schaeffer. An experiment to measure the usability of parallel programming systems. *Concurrency: Practice and Experience*, 8(2):147–166, 1996.
13. Andrew Witkin and Michael Kass. Reaction–diffusion textures. *Computer Graphics (SIGGRAPH '91 Proccedings)*, 25(4):299–308, July 1991.
14. A. Zubiri. An assessment of Java/RMI for object–oriented parallelism. Master's thesis, Department of Computing Science, University of Alberta, 1997.

The IceT Environment for Parallel and Distributed Computing *

Paul A. Gray and Vaidy S. Sunderam

Emory University

Abstract. The current programming models, tools and environments associated with Internet programming and parallel, high-performance distributed computing have remained isolated from one another. The focus of the IceT project has been to bring together the common and unique attributes of these areas. The result is a confluence of technologies in a parallel programming environment with several novel characteristics. The result provides users with a parallel, multi-user, distributed programming environment; upon which processes and data are allowed to migrate and to be transfered throughout owned and unowned resources, under security measures imposed by owners of the local resources.

1 Introduction

Distributed computing over locally networked workstations has reached a solid level of maturity in recent years. Coupled with recent and considerable gains in processor performance and network capabilities, software tools, programming languages and methodologies, parallel environments consisting of clustered workstations remain viable platforms for high-performance computation [7, 11]. New developments in these and related areas have been the motivating factor behind the development of IceT. IceT builds upon traditional distributed computing techniques and paradigms to include novel aspects which have been brought about by these evolving technologies.

The IceT resource model consists of multiple clusters of virtual environments belonging to multiple users, each with distinct levels of security and accessibility. The resulting collocation of resources comprises a multi-user, multi-level, time-shared virtual machine which provides and facilitates interaction and collaboration among users and processes. Fundamental to the process and data model of IceT is the property of transportability. Consequently, native IceT processes and data are fluid entities which may be uploaded to remote locations and, in the case of processes, be executed without regard for the remote architecture or file system structure. For a more detailed description of the components of IceT, see [6]. By combining multiple users, their respective resource pools, and allowing portability and uploadability of tasks and data, IceT provides a new genre

* Research supported by ARO grant DAAH04-96-1-0083, DOE Grant No. DE-FG05-91ER25105, NASA Grant NAG 2-828 and NSF Award No. ASC-9527186.

for distributed computing in areas such as collaborative computing and active agents, and significantly extends the possibilities for numerical computations.

The remaining portion of this paper provides an overview of the IceT system. Section 2 presents the user-level configurability and registration of the virtual environments. Section 3 provides an example of how one might use the IceT substrate for programming collaborative tools, and Sec. 4 contrasts IceT's performance with the established distributed programming environment of PVM [4].

2 IceT's Virtual Machine Configuration

In many ways, IceT's distributed and parallel programming protocols are modeled after PVM. Users can *add*, *delete*, and otherwise configure the resources within their virtual environments and can *spawn*, *kill* and likewise manipulate processes within the environment. In addition to the traditional paradigm, users are able to *merge* with existing virtual environments of others and *withdraw* (intact) components and sub-environments.

The IceT objectives include enabling a *nomadic* environment which can collect, release, and abandon subsystems on which processes flow freely among owned and unowned resources. As a consequence, a collection of distributed processes can dynamically manipulate, conform to, and adapt to its computational foundation. For example, a multi-stage process consisting of computations, data manipulation, and subsequent visualization of results would be able to start up on a cluster of high-performance workstations for the computations; upon completion, migrate with the data to a mainframe where the data from the computations would be archived and parsed; and finally, the data of interest could move to a cluster of 3-D graphics workstations for visualization. Additionally, in combining these nomadic characteristics of IceT, the support for the mobile processes of IceT as described above has many implications to fault tolerance and load balancing.

The manipulation of the environment may be performed within an IceT program or by an IceT "console" (currently a text-based console or through a graphical user interface to the console). Figure 1 shows a sample user interaction with the text-based IceT console. In this example, a user constructs a local IceT subsystem consisting of the computational components sparc1, sparc2, sparc3, and cauchy (lines 1-9). Subsequently, this user merges the local system with the external resources and subsystems known to ntpc (line 10). After the merge, the configuration of the virtual machine consists of the entities enrolled by the local user and the entities which were currently enrolled with the external resource (shown in lines 11-28). Communication routes to non-owned resources are dictated by the owner of the resource. In this example, ntpc provides access to four additional resources (lines 17-24), but regulates the communication channels and obfuscates their Internet addresses and ports. Users and processes can then spawn processes on any of the computational components of the environment (lines 29-31). The -h flag supplied to the spawn request (line 29) is a "host" flag where the user can dictate the host on which to execute the process.

```
01:gray@cauchy:~> java IceT.Console
02:
03:- IceT%> add sparc1 sparc2 sparc3
04:- IceT%> conf
05:      cauchy, id #47153 (Daemon)
06:      sparc1, id #33960 (Daemon)
07:      sparc2, id #34170 (Daemon)
08:      sparc3, id #33156 (Daemon)
09:      cauchy, id #47150 (Console)
10:- IceT%> merge ntpc.mathcs.emory.edu 1079
11:- IceT%> conf
12:      cauchy, id #47153 (Daemon)
13:      sparc1, id #33960 (Daemon)
14:      sparc2, id #34170 (Daemon)
15:      sparc3, id #33156 (Daemon)
16:      ntpc.mathcs.emory.edu, id #1079 (Daemon)
17:      resource1, (Daemon)
18:         (routed through ntpc.mathcs.emory.edu, id #1079 (Daemon))
19:      resource2, (Daemon)
20:         (routed through ntpc.mathcs.emory.edu, id #1079 (Daemon))
21:      resource3, (Daemon)
22:         (routed through ntpc.mathcs.emory.edu, id #1079 (Daemon))
23:      resource4, (Daemon)
24:         (routed through ntpc.mathcs.emory.edu, id #1079 (Daemon))
25:      cauchy, id #47150 (Console)
26:      ntpc.mathcs.emory.edu, id #1087 (Console)
27:      resource3, (Console)
28:         (routed through ntpc.mathcs.emory.edu, id #1079 (Daemon))
29:- IceT%> spawn MatrixMultiply -h resource3
30:  Waiting for task response
31:  Task spawned -> resource3, (Task)
32:         (routed through ntpc.mathcs.emory.edu, id #1079 (Daemon))
33:- IceT%>
```

Fig. 1. Sample user interaction with IceT console.

A "merge" request requires some additional knowledge of external resources. This information could easily be obtained in one of several ways. Currently, such a server is implemented in the form of a cgi-bin script controlled by an http server. An IceT daemon can "register" the cluster by submitting a form to a simple Perl script in the http server's cgi-bin directory, which appends this information to a database. An IceT daemon can "request" additional resources by requesting the database (in html format) from the http server and examining the entries in the database for suitable resources.

3 Collaborative Computing with IceT

In bridging the gap between distributed virtual environments, not only does one look to reap benefits from the increase in computational resources, but also to the natural collaborative and distributed environment which results. This section introduces the collaborative and distributing computing aspects supported by IceT.

IceT provides the framework on which collaborative tools can be built; this is in contrast to other ongoing large-scale web-based collaborative projects (see [2] and the references cited therein). In IceT, processes and users of the processes are based upon a more liberal paradigm, in the sense that users are allowed to access, execute and migrate processes on arbitrary resources in the established environment. Moreover, processes are able to interact with multiple users. For example, the owner of a collaborative tool can *just-in-time-install*, or *soft-install*, a process upon colleagues' resources.[1] The colleagues can join in the collaboration and interact with the tool.

Several novel features incorporated into IceT are illustrated by way of a simple IceT-based "chat" tool (Fig. 2). In this example, the `IceTChat` tool is owned by a single user in the sense that the static form of the process resides on the user's local filesystem. This user wishes to share the tool with other users involved in the collaboration. The setting for this example is an IceT environment consisting of several merged IceT users' environments (perhaps established in the manner depicted in Sec. 2).

The users of the aggregate IceT environment are engaged in a collaborative session of some sort, perhaps a joint visualization or computational simulation. One of the users would like to "chat" with the other participants about the goings-on in the simulation and gain a consensus on how to proceed. To do so, this user starts up an IceTChat tool which resides on the local system. The IceTChat tool queries the IceT environment for a list of users who may be brought in to the IceTChat session. IceT users are found on the machines `katana`, `electron` and `nirvana`.

At this point, the user specifies a host to chat with, and the IceTChat tool spawns the IceTChat process on the remote host. If the spawn request fails for some reason, an `IceTSpawnException` is thrown by the spawn routine, and the user is notified of the failed spawn request. The following code shows the corresponding `addParticipant` syntax taken from the main `IceTChat.java` source file:

---

[1] Upon termination of processes which have been "uploaded" to remote hosts, the classes associated with the process are garbage-collected by the Java Virtual Machine and no longer reside on the remote system. Thus, the "soft-installation" terminology alludes to this feature of just-in-time installations.

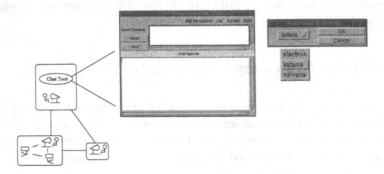

Fig. 2. Starting IceTChat tool, identifying possible participants.

```
public synchronized void addParticipant(String hostToChatWith) {
    try {
        IceT_spawn("IceTChat",idFor(hostToChatWith));
    }
    catch (IceTSpawnException itse) {
        notify("Chat Request Failed: "+itse.toString());
    }
}
```

The IceTChat's spawn request is forwarded to the IceT daemon on the specified host. The `createTask` code below is internal to the IceT substrate (i.e., is not accessible to a user's program) and is called by an IceT daemon to create a new instance of a process. Upon invocation of `createTask`, the task and it's dependency classes are located, possibly on a remote host. The "spawning" host is the default remote host to seek class files not found on the local resource.

```
protected static synchronized void createTask(String task,
    Id Spawner, String params)
    throws IceTException {

    int flag = locateAndResolve(task,Spawner);
```

            (Remaining portion of `createTask` omitted.)

The `locateAndResolve` function's actions are highly dependent on the security policies imposed upon the local resource. Without getting into burdensome details, the simplest result is to locate the main class file named `task`, and to register, load, and resolve dependencies of the class using IceT's `ClassBootStrapper`. The `ClassBootStrapper` may be likened to a Web browser, which uploads and instantiates classes (and packages) required by the primary class.

Once all of the package and class files that make up the the complete IceTChat tool have been uploaded to the remote host, the IceTChat tool may finally be instantiated as a separate process or as a thread, governed by the remote IceT daemon and the security policies. In short, a *just-in-time* installation is performed by the `ClassBootStrapper` and the IceTChat tool has been *soft-installed* on the remote host.

The `IceTChat` program begins execution on the remote machines, gives notice to the local user that a chat session is starting up and allows the user to join in the session (Fig. 3). Finally, with all of the IceTChat tools on board, the IceTChat

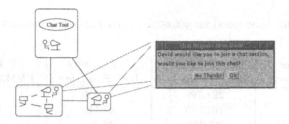

Fig. 3. IceTChat tool asks for participation.

session uses the IceT message-passing substrate to pass messages among the enrolled entities (Fig. 4). When the users finish their IceTChat session, they

Fig. 4. IceTChat uses IceT's message-passing facilities.

may save uploaded IceTChat tool classes locally or have all of the components removed from their system (via garbage-collection).

Although "chat" programs are commonplace, the support for platform independence and the portability of processes by IceT shown here demonstrate potential for the development of far more complex collaborative tools. One could easily evolve the simple chat tool to collaborative tools for joint visualizations, data manipulation, or distributed instruction.

4 Performance Comparisons

The status of Java as a scientific programming language has been the subject of debate since Java's early release. Consequently, the question of performance of such a system which depends on Java is quite valid. Recent remarks which address these and other issues can be found in [3,1].

Table 1 gives some indication of how the IceT programming environment compares with PVM implementations for the same problem, a simple block-decomposed matrix-matrix multiplication. A full characterization of the Java-binding of the IceT implementation, including program listings, is given in [5].

Table 1. Compute time (secs) for 400x400 matrix multiplication on a SPARC20 cluster.

# Processes (Master+Slaves)	Java Processes IceT Substrate	C Processes IceT Substrate	C Processes PVM3.3 Substrate
1	291.796	41.480	37.134
2	161.589	45.116	24.608
4	87.044	28.920	16.575
8	50.695	21.287	16.272

The first column of times are those used by Java-based Master and Slave programs supported by the IceT substrate. The next column shows the performance when the same IceT substrate is used for task creation and communication, but the actual computations are performed using C-based Master and Slave programs. This result shows that IceT, in its evolving utility capacity, can serve as a conduit through which system-*dependent* code can be ported. So computationally-intense parallel programs already written in C or FORTRAN (to obtain significant performance gains over Java, as seen when comparing the first two columns of times above) could be ported among appropriate architectures within the extended IceT environment. The last column shows times for the equivalent C-based Master and Slave programs executed under PVM 3.3. The last two columns show that IceT mixed with system-dependent computational components performs quite competitively with PVM.

5 Future of IceT

Recent improvements to the JDK-1.1 (JNI [8], RMI [10] and Object Serialization [9] to name a few) have had positive impacts on portability and access of processes within IceT. Developments in the areas of active agents and active networks using IceT are also being pursued, as are issues associated with security, performance, load balancing, fault-tolerance, and the far-reaching implications of nomadic virtual machines.

References

1. B. Carpenter, Y.-J. Chang, G. Fox, D. Leskiw, and X. Li. Experiments with 'HP Java'. *Concurrency, Experience and Practice*, 9(6):633–648, June 1997.

282

2. M. Chen and J. Cowie. Java's role in distributed collaboration. *Concurrency, Experience and Practice*, 9(6):509–520, June 1997.

3. G. C. Fox and W. Furmanski. Java for parallel computing and as a general language for scientific and engineering simulation and modeling. *Concurrency, Experience and Practice*, 9(6):415–426, June 1997.

4. G. A. Geist and V. S. Sunderam. The PVM system: Supercomputer level concurrent computation on a heterogeneous network of workstations. In *Proceedings of the Sixth Distributed Memory Computing Conference*, pages 258–261. IEEE, 1991.

5. P. Gray and V. Sunderam. IceT: Distributed computing and Java. *Concurrency, Experience and Practice*, November 1997.

6. P. Gray and V. Sunderam. The IceT framework for metacomputing. Source available at http://www.mathcs.emory.edu/~gray/IceT3.ps, 1997.

7. J. Salmon and M. S. Warren. Parallel out-of-core methods for n-body simulation. In *8th SIAM Conference on Parallel Processing for Scientific Computing*, Philadelphia, 1997. SIAM.

8. Sun Microsystems. Java Native Method Invocation Specification. Source available at ftp://www.javasoft.com/docs/jdk1.1/jni.pdf, May 1997. Revision 1.1.

9. Sun Microsystems. Java Object Serialization Specification. Source available at ftp://www.javasoft.com/docs/jdk1.1/serial-spec.pdf, February 1997. Revision 1.3.

10. Sun Microsystems. Java Remote Method Invocation Specification. Source available at ftp://www.javasoft.com/docs/jdk1.1/rmi-spec.pdf, February 1997. Revision 1.4.

11. M. S. Warren, D. J. Becker, M. P. Goda, and T. Sterling. Parallel supercomputing with commodity components. In H. R. Arabnia, editor, *Proceedings of the International Conference on Parallel and Distributed Processing Techniques and Applications*, pages 1372–1387, 1997.

A General Resource Reservation Framework for Scientific Computing*

Ravi Ramamoorthi, Adam Rifkin, Boris Dimitrov, and K. Mani Chandy

California Institute of Technology

Abstract. We describe three contributions for distributed resource allocation in scientific applications. First, we present an *abstract model* in which different resources are represented as tokens of different colors; processes acquire resources by acquiring these tokens. Second, we present *distributed scheduling algorithms* that allow multiple resource managers to determine custom policies to control allocation of the tokens representing their particular resources. These algorithms allow multiple resource managers, each with its own resource management policy, to collaborate in providing resources for the whole system. Third, we present an implementation of a distributed resource scheduling algorithm framework using our abstract model. This implementation uses Infospheres, which are Internet communication packages written in Java, and shows the benefits of distributing the task of resource allocation to multiple resource managers.

1 Introduction

A user often needs access to several distributed heterogeneous resources. For instance, a scientist may conduct a distributed experiment [3] requiring a supercomputer, a visualization unit, and a special high quality printer all in different locations. All three resources are essential to the experiment so the scientist needs to *synchronously* lock and use all three *distributed resources* for the same time period to complete the computing task. The distributed heterogeneous resources together form a *networked virtual supercomputer* or *metacomputer* [1]. The scientist also wants resources to be scheduled automatically as a service of the appropriate software, with or without the inclusion of specific supplemental information such as the times the user is available to perform the experiment.

Traditional metacomputing resource allocation [6, 9] uses a central authority for scheduling, usually for efficiency. For example, the IBM SP2 uses a scheduling algorithm [8] that reduces the wait time of jobs requiring only a few nodes, if these can be scheduled without delaying more computationally intensive jobs.

* This work was supported in part under the Caltech Infospheres Project, by the Air Force Office of Scientific Research under grant AFOSR F49620-94-1-0244, by the CISE directorate of the NSF under Problem Solving Environments grant CCR-9527130, and by the NSF Center for Research on Parallel Computation under cooperative agreement CCR-9120008. We thank Doug Lea for his helpful comments.

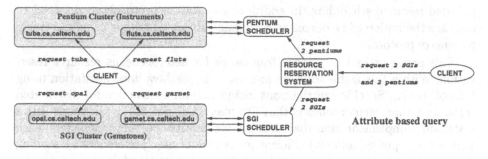

Fig. 1. Two models are given for resource reservation. On the left, the client simply asks for specific machines. On the right is a more advanced request, in which the client asks the Resource Reservation System (RRS) for two SGIs and two Pentiums. The RRS connects to separate resource managers that schedule time on the two clusters (using, for example, our calendar-based algorithm).

By contrast, consider the computational needs of users requiring resources managed by different groups in different places. Scheduling is more complicated because it is impractical for individual sites to "know" global information that would help them to do more efficient scheduling [6].

The owner of a set of resources may have resource management policies that are different from those of owners of other resource sets. Our challenge is two-fold: (i) to establish methods of cooperation so that the collection of owners offers system-wide resources to users, and (ii) to make the algorithms scalable so that new resource providers can enter the common resource pool quickly and semi-autonomously.

An infrastructure for reserving resources in a distributed system is required by many applications. Our research deals with designs and implementations of distributed resource management schemes that coordinate different policies for different sets of resources. Though this paper addresses resources used in meta-computing, our research deals with resources in many distributed applications.

A convenient abstraction for such applications represents each indivisible resource by an indivisible *token* of some color [4]; different types of resources have different colors. For instance, a node of an IBM SP2 can be represented as a token of the IBM SP color. Likewise, a room in a hotel can be represented by a token of the hotel color.

Our model deals with time explicitly. So, a reservation can be made for 64 nodes of an IBM SP2 for 10 contiguous hours, or a hotel for seven nights.

The centralized IBM SP2 scheduling algorithm relies on knowledge of how many nodes each process needs to "promote" less computationally-intensive tasks as necessary. On the other hand, as illustrated in Fig. 1, if each node in a supercomputer were to be scheduled independently in a distributed way, efficient scheduling would become much more difficult. As metacomputing applications use distributed heterogenous systems, they will need algorithms for efficient distributed resource scheduling. In addition, negotiation protocols might need to leverage the notion of resources as economic currency, perhaps using electronic commerce protocols.

This paper presents a general framework for heterogeneous resource reservation. Within this framework, we present a simple Java implementation using Infospheres [2]. Specific contributions include: (i) an abstraction for distributed resource management problems that fits many, but not all, applications; (ii) a distributed implementation that coordinates multiple resource managers, each with its own policy; and (iii) efficient processing of user preferences by sending Java applets to resource managers to perform resource scheduling.

In Sec. 2, we discuss some simple attempts at distributed resource allocation algorithms, describe how they fail, and introduce *calendars*, which are useful for efficient resource allocation. In Sec. 3, we describe how the calendar metaphor builds on our resources-as-tokens metaphor. A simple application to safe metacomputer scheduling across distributed resource managers is presented in Sec. 4, after which we discuss efficient scheduling when resources are specified by attribute. We conclude with some observations in Sec. 5.

2 Distributed Resource Reservation Algorithms

The problem of distributed resource reservation has several simple solutions, including local clocks and central server, that are correct but may be inefficient. We discuss how calendars provide a more scalable solution.

One approach to resource reservation is to try to lock all of the resources the application wants. If an application is unable to lock a resource, it enters a queue waiting for it based on the priority of a logical *local clock* timestamp [4]. If an application with lower priority has the resource but is not yet using it, that application must relinquish the resource (or token), deferring to the higher-priority application. This method is robust and fairly scalable, but can be inefficient. For example, as illustrated in Fig. 2, client 1 can be using resource 1, while client 2 is waiting to use it. Client 2 has locked resource 2 and is not using it, but still prevents client 3 from using 2 (which 3 could use since it requires no other resources to run its task).

As discussed in Sec. 1, we could improve efficiency by using a *central scheduling* algorithm, as used by the IBM SP2. However, this is clearly impractical from a scalability standpoint. Our goal is to recover some or all of the efficiency of a worldwide central server while maintaining the scalability features of distributed resource-management servers.

Calendars allow a nice tradeoff between scalability of resource managers and efficient utilization of resources. Allowing an application to "make appointments" in a calendar for resource reservation, *a resource cannot be blocked from use while sitting idle*. If a resource is unused, no application has an appointment for it at that time. Thus, efficient resource allocation is possible without global information. This calendar model is easily extensible to general resource reservation.

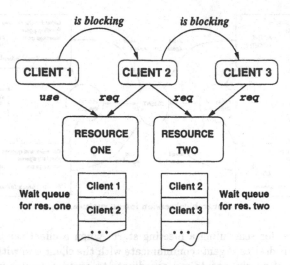

Inefficient local clocks solution.

Fig. 2. Local clocks can fail to use available resources. Client 1 holds resource 1, and client 2 is next in line for both resources 1 and 2. Because client 2 is blocking it, client 3 must wait for client 2 to finish (2 has higher priority), and hence client 1 to finish, even though client 3 does not even use resource 1.

3 Our Model

Individual resources use a calendar metaphor for arranging their schedule; the basic calendar functionality our implementation provides includes the concepts of time slots and access lists.

A *time slot* consists of a time interval with a particular time unit grain. Every time slot can be in one of three states: *locked, held* or *available*. A *Locked* slot appears when a client commits to using a resource during that slot; as a result, locked slots can be read but not written. *Held* slots are slots that a particular client is considering locking, but has not yet committed to locking. Only that client can write to these slots, thereby locking them; they are read-only for other clients. However, unlike a locked slot, a held slot can be released, reverting to *available*. Available slots may be read or written and converted to *held* or *locked* status. Slots correspond to the tokens discussed in Sec. 1; so, resource reservation is tantamount to collecting the proper tokens.

Each slot has an associated *access list* that keeps track of which processes can obtain a lock on that slot. For instance, a resource manager may provide access to an authorized user from the Center for Research on Parallel Computation, but not grant access to anyone else. Thus, some slots may be available to only one set of users, while others can be available to other sets of users. This approach differs from traditional "whiteboard scheduling" models.

The *reservation* of a set of resources is determined when all of the resource managers (or servers) agree to lock the slots that correspond to the same time.

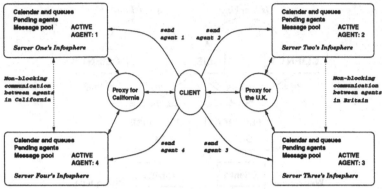

Hierarchical session infrastructure.

Fig. 3. Our model for scheduling a meeting starts when a client sends agents to the various resources it desires. Agents communicate with the client and with resources. For efficiency, groups of nearby agents can coordinate to avoid excessive message-passing to clients, who may be geographically distant. The resource managers or servers can also send agents to clients to request back the slots that the clients hold.

We implement reservations atomically using a two-phase commit protocol [7]. The action starting with resource-request initiation and ending with resource-reservation commitment corresponds to an Infosphere session [5].

Reserving Resources. Our paradigm for resource reservation, using client requests and brokering agents, provides a test bed on which effective algorithms can be developed for specific tasks; see Fig. 3. Resource reservation begins with a client application making a request, the only interaction a user needs to have with the system. A resource can be represented by a boolean function over all possible Cartesian products of resources and meeting times, with additional weights given to represent hints. For example, requiring one Pentium and one SGI on Monday at 10AM is a request that assigns boolean **true** to all combinations of resources that include the desired Pentium and SGI. In addition, hints can help the system choose more appropriate scheduling policies. Although the general framework is too complex to implement directly in some applications, for any particular application a suitable subset can be implemented.

Like ambassadors to foreign countries, the client system can send a small set of instructions in Java as *agents* [10] to any resource manager to request computing time. Several efficiency improvements make agent communication attractive. Agents can include user preferences for efficient filtering of available times at the server end. The filtered set can then be returned to the client, thereby avoiding heavy message passing in congested or high latency networks. Since nearby agents can designate a common agent to efficiently set up a coordinated reservation time among these agents, hierarchical solutions can be used to obtain lists of available slots. By varying the programs that the agents define, different algorithms can easily be tested without major modifications in the system.

Not only can clients send agents to servers, but servers can send agents to clients to request back slots that clients had on hold, upon request from a client that has higher priority. Our system requires that the agent recipient must lock the slot withing a time period, or the slot will be automatically returned to the resource manager.

Agent Primitives. Scheduling agents communicate with resource managers on servers using query, lock, release, and wait messages (Fig. 4).

Queries. A query is the first communication an agent makes when setting up a meeting. When a server receives a query, it gives the client's agent complete (access-dependent) information about which slots are available, held, and locked. The agent relays (a possibly filtered version of) this information back to the client. However, it also executes quickly on the server, filtering this information to reserve some vacant slots and wait for its client to decide what to do with them. The server may impose restrictions on how many slots the agent can reserve at any one time. We could dispense with agents and allow the server to pick the slots it reserves for the client; although this is our implementation default, the agent innovation allows the client to encode some preference information and have it honored without the lag of message-passing.

Locks and Releases. The server allows authorized clients to lock slots they hold or release uncommitted slots. It sends released slots to the highest-priority client on the waiting list if one exists.

Waits. The server can receive requests to be placed on the waiting list for specific slots. If the slot is held, but not committed, the server will honor the request and, if the requester has higher priority than the current holder, the server will request the current holder to return the slot. The holder must lock the slot within a certain time period, or it will be returned automatically.

4 Applications

Two applications illustrate our framework: scheduling specific resources controlled by more than one resource manager, and scheduling by attribute.

Multiple Resource Managers. Consider scheduling two or more resources, each controlled by a different manager. One solution is to use local clocks (discussed in Sec. 2) to place on hold each resource's calendar before scheduling computing time by locking the appropriate slots. That has the efficiency problem discussed in Sec. 2, but it will be smaller since we are using the algorithm only to schedule calendars, not resource use. Thus, just introducing the calendar metaphor provides substantial savings.

In this algorithm, when one user is reserving time on a given resource, all other users are excluded, while in reality we need mutual exclusion only on individual slots for safety. We can therefore improve this algorithm's efficiency: a client can use finer-grained adaptive control to place on hold only a small part of the resource manager's calendar at a time.

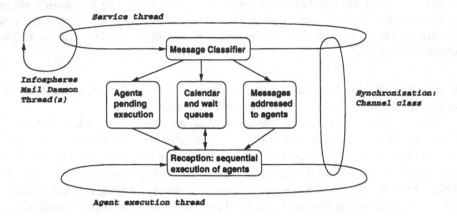

Fig. 4. Agents execute atomically and communicate with their server and the outside world via a receptionist class which provides only non-blocking send and receive methods. Though agents run in the address space of the server, privacy is possible via the Java security manager.

Query. The client indicates interest in scheduling a resource by sending agents to the various servers. The agent executes the program given to it by the client, and after communicating with the resource manager returns to the client's system a list of slots that it is currently holding, and information (which in case of free slots may not be up to date) on whether the remaining slots are free, on hold, or locked. Note that the server pipes available slots through the agent giving the client a small number of desirable held slots. The agent is an efficient way to encode client preferences cheaply.

Schedule if possible. The client can then schedule computing time if at least one slot returned to it from all of the resources matches. It writes to that slot, committing to using all of the resources at that time, makes any necessary payments for use of those resources, and releases the remaining slots.

Negotiate. If the client cannot immediately schedule all required resources, it negotiates instead of just giving up and retrying. Specifically, it releases all held slots that were locked by other users in at least one desired resource, since there is no chance of getting all desired resources for that time slot. It then enters a queue on other slots, where there is still a chance of acquiring all desired resources. Using logical clocks, the algorithm continues until resources are reserved, or no reservation is possible. The negotiation phase is usually unnecessary.

Change the agent. Based on the type of negotiation required, the client can keep evolving its agent to better meet its changing needs for placing holds as well as locking and releasing slots.

Resource Reservation by Attribute. Offering reservation by attribute (for example, a request for "3 SGIs and 2 Pentiums") is easily integrated into our existing framework. The "and" clause defines resources that must be reserved together, so these can be treated as specific resources themselves. This reduces to a scheduling problem such as "get 3 SGIs out of the 30 known to my system." We want n out of k homogeneous resources where $(k > n)$. We can use the algorithm for multiple resource managers if we send agents to p out of the k resources, but then pick the "best" (or earliest) time at which n out of the p polled resources are available. Our problem then reduces to choosing p; choosing $(p = k)$ may not always be the best solution due to message passing delay. For this reason, we have developed a simple mathematical model for choosing the optimal p. In our model, the expected *delay* in scheduling a job is computed for each p using the probability q that a given slot is unavailable. We then plot a graph of cost versus p to find the p that minimizes the delay.

5 Conclusions

We have investigated generalizable resource allocation algorithms for which desired resources can be specified by attribute only, and for which different resource managers can coordinate synchronously. Our model builds on the concept of *resources as tokens* and the metaphor of *calendars for scheduling*. To improve efficiency under high network latency, our implementation passes small Java programs as agents for coordination. Our design represents the first step toward the development of a robust scheduling infrastructure, layered above conventional schedulers currently available, for the next generation of virtual supercomputers constructed from heterogeneous resources distributed over the Internet.

References

1. C. Catlett and L. Smarr. Metacomputing. *Comm. of the ACM*, 35:44–52, 1992.
2. K.M. Chandy, J. Kiniry, A. Rifkin, and D. Zimmerman. A framework for structured distributed object computing. *Parallel Computing*, 1997. Submitted.
3. K.M. Chandy, J. Kiniry, A. Rifkin, and D. Zimmerman. Webs of archived distributed computations for collaboration. *Journal of Supercomputing*, 11(1), 1997.
4. K.M. Chandy and J. Misra. *Parallel Program Design*. Addison-Wesley, 1988.
5. K.M. Chandy and A. Rifkin. Systematic composition of objects in distributed internet applications: Processes and sessions. *Proceedings of the Thirtieth Hawaii International Conf. on System Sciences*, pages 395–404, January 1997.
6. I. Foster and C. Kesselman. Globus: A metacomputing infrastructure toolkit. *International Journal of Supercomputer Applications*, 1997.
7. J. Gray and A. Reuter. *Transaction Processing*. Morgan-Kaufmann, 1993.
8. D.A. Lifka, M.W. Henderson, and K. Rayl. Users guide to the argonne sp scheduling system. Technical Report ANL/MCS-TM-201, Argonne, May 1995.
9. M. Litzkow, M. Livney, and M. Mutka. Condor – a hunter of idle workstations. In 8th *International Conf. on Distributed Computing Systems*, pages 104–111, 1988.
10. P. Maes. Agents that reduce work. *Comm. of the ACM*, 37(7):31–40, July 1994.

Author Contacts

Ju-Pin Ang, DSO National Laboratories, 20 Science Park Drive, Singapore 118230, jpang@singnet.com.sg.

Scott B. Baden, Computer Science and Engineering Dept., U. California at San Diego, 9500 Gilman Drive, La Jolla, CA 92093-0114, USA, baden@cs.ucsd.edu.

Donald Bashford, Department of Molecular Biology, The Scripps Research Institute, 10550 N. Torrey Pines Road, La Jolla, CA 92037, USA, bashford@scripps.edu.

Federico Bassetti, Scientific Computing Group CIC-19, Los Alamos National Laboratory, Los Alamos, NM 87545, USA, fede@lanl.gov.

Matthias Besch, RWCP Massively Parallel Systems GMD Laboratory Berlin, Rudower Chaussee 5, D-12489 Berlin, Germany, mb@first.gmd.de.

Hua Bi, RWCP Massively Parallel Systems GMD Laboratory Berlin, Rudower Chaussee 5, D-12489 Berlin, Germany, bi@first.gmd.de.

Robert A. Bond, MIT Lincoln Laboratory, Lexington, MA 02173, USA.

Are Magnus Bruaset, Numerical Objects AS, P.O. Box 124, Blindern, N-0314 Oslo, Norway, amb@nobjects.com.

Xin Cai, Department of Informatics, University of Oslo, P.O. Box 1080, Blindern, N-0316 Oslo, Norway, xingca@ifi.uio.no.

C. Calvin, CEA Grenoble, 17 Av. des Martyrs, 38054 Grenoble cedex 9, France, calvin@alpes.cea.fr.

K. Mani Chandy, Department of Computer Science, MS 256-80, California Institute of Technology, Pasadena, CA 91125, USA, mani@cs.caltech.edu.

Andrew A. Chien, Departmentof Computer Science, University of Illinois at Urbana-Champaign, Urbana, IL 61801, USA, concert@red-herring.cs.uiuc.edu.

Julian C. Cummings, Advanced Computing Lab, Los Alamos National Laboratory, Los Alamos, NM 87545, USA, julianc@acl.lanl.gov.

Jim M. Daly, MIT Lincoln Laboratory, Lexington, MA 02173, USA.

Kei Davis, Scientific Computing Group CIC-19, Los Alamos National Laboratory, Los Alamos, NM 87545, USA, kei@lanl.gov.

Cecelia DeLuca, MIT Lincoln Laboratory, Lexington, MA 02173, USA.

Boris Dimitrov, Department of Computer Science, MS 256-80, California Institute of Technology, Pasadena, CA 91125, USA, boris@cs.caltech.edu.

Julian Dolby, Departmentof Computer Science, University of Illinois at Urbana-Champaign, Urbana, IL 61801, USA, concert@red-herring.cs.uiuc.edu.

Ph. Emonot, CEA Grenoble, 17 Av. des Martyrs, 38054 Grenoble cedex 9, France, emonot@alpes.cea.fr.

Stephen J. Fink, Computer Science and Engineering Dept., U. California at San Diego, 9500 Gilman Drive, La Jolla, CA 92093-0114, USA, sfink@cs.ucsd.edu.

Nobuhisa Fujinami, Sony Computer Science Laboratory Inc., Takanawa Muse Building, 3-14-13 Higashi-Gotanda, Shinagawa-ku, Tokyo 141, Japan, fnami@csl.sony.co.jp.

Bishwaroop Ganguly, Departmentof Computer Science, University of Illinois at Urbana-Champaign, Urbana, IL 61801, USA, concert@red-herring.cs.uiuc.edu.

Jens Gerlach, Tsukuba Research Center of RWCP, Tsukuba Mitsui Building, 1-6-1 Takezono, Tsukuba-shi 305, Japan.

Vladimir Getov, School of Computer Science, University of Westminster, Harrow Campus, Horthwick Park, Harrow HA1 3TP, UK, v.s.getov@westminster.ac.uk.

John Gilbert, Xerox PARC, 3333 Coyote Hill Road, Palo Alto, CA 94304, USA, gilbert@parc.xerox.com.

Paul A. Gray, Mathematics and Computer Science Department, Emory University, 1784 N. Decatur Road, Atlanta, GA 30322, USA, gray@mathcs.emory.edu.

Russell F. Haddleton, Department of Computer Science, University of Virginia, Charlottesville, VA 22903, USA, rfh2y@cs.virginia.edu.

Gerd Heber, RWCP Massively Parallel Systems GMD Laboratory Berlin, Rudower Chaussee 5, D-12489 Berlin, Germany, heber@first.gmd.de.

Curtis W. Heisey, MIT Lincoln Laboratory, Lexington, MA 02173, USA.

Michael E. Henderson, IBM T.J. Watson Research Center, P.O. Box 218, Yorktown Heights, NY 10598, USA, mhender@watson.ibm.com.

William Humphrey, Advanced Computing Lab, Los Alamos National Laboratory, Los Alamos, NM 87545, USA, bfh@acl.lanl.gov.

Yuuji Ichisugi, Electrotechnical Laboratory, Japan, ichisugi@etl.go.jp.

John Irwin, Xerox PARC, 3333 Coyote Hill Road, Palo Alto, CA 94304, USA.

Tutaka Ishikawa, Real World Computing Partnership, Tsukuba Mitsui Building 16F, 1-6-1 Takezono, Tsukuba-shi, Ibaraki 305, Japan, ishikawa@trc.rwcp.or.jp.

Jaakko Järvi, Turku Centre for Computer Science, Lemminkäisenkatu 14, FIN-20520 Turku, Finland, jjarvi@cs.utu.fi.

Wei-Min Jeng, Department of Computer Science, University of Houston, Houston, TX 77204, USA, pet@cs.uh.edu.

M. Ed Jernigan, Department of Systems Design Engineering, University of Waterloo, Waterloo, Ontario, Canada N2L 3G1.

Wouter Joosen, K. U. Leuven, Department of Computer Science, Celestijnenlaan 200A, B-3001 Leuven, Belgium.

Vijay Karamcheti, Department of Computer Science, University of Illinois at Urbana-Champaign, Urbana, IL 61801, USA, concert@red-herring.cs.uiuc.edu.

Steve Karmesin, Advanced Computing Lab, Los Alamos National Laboratory, Los Alamos, NM 87545, USA, karmesin@acl.lanl.gov.

Noel D. Keen, University of New Mexico, Albuquerque, NM 87131, USA, noelk@unm.edu.

Matthias Kessler, RWCP Massively Parallel Systems GMD Laboratory Berlin, Rudower Chaussee 5, D-12489 Berlin, Germany, mk@first.gmd.de.

Gregor Kiczales, Xerox PARC, 3333 Coyote Hill Road, Palo Alto, CA 94304, USA, gregor@parc.xerox.com.

John Lamping, Xerox PARC, 3333 Coyote Hill Road, Palo Alto, CA 94304, USA, lamping@parc.xerox.com.

Hans Petter Langtangen, Mechanics Division, Department of Mathematics, University of Oslo, P.O. Box 1080, Blindern, N-0316 Oslo, Norway, hpl@math.uio.no.

Stephen R. Lee, X-CI MS F663, Los Alamos National Laboratory, Los Alamos, NM 87545, USA, srlee@lanl.gov.

Camilliam Lin, Anadrill, Schlumberger Oil and Well Service, Sugar Land, TX 77478, USA, clin@cugar-land.anadrill.slb.com.

Gerald Löffler, Research Institute of Molecular Pathology, Dr. Bohr-Gasse 7, A-1030 Wien, Austria, Gerald.Leoffler@univie.ac.at.

Jean-Marc Loingtier, Xerox PARC, 3333 Coyote Hill Road, Palo Alto, CA 94304, USA.

Andrew Lumsdaine, Laboratory for Scientific Computing, Department of Computer Science and Engineering, University of Notre Dame, Notre Dame, IN 46556, USA, lums@lsc.nd.edu.

Stephen L. Lyons, Mobil Technology Company, 13777 Midway Road, Dallas, TX 75244, USA, sllyons@dal.mobil.com.

Steve MacDonald, Computer Science Department, University of Alberta, Edmonton, Alberta, Canada, T6G 2H1, stevem@cs.ualberta.ca.

Motohiko Matsuda, Real World Computing Partnership, Tsukuba Mitsui Building 16F, 1-6-1 Takezono, Tsukuba-shi, Ibaraki 305, Japan, matu@trc.rwcp.or.jp.

S. Matsuoka, Department of Information Science, Tokyo Institute of Technology, Tokyo, Japan, matsu@is.titech.ac.jp.

Brian McCandless, Laboratory for Scientific Computing, Department of Computer Science and Engineering, University of Notre Dame, Notre Dame, IN 46556, USA, McCandless.1@nd.edu.

Anurag Mendhekar, Xerox PARC, 3333 Coyote Hill Road, Palo Alto, CA 94304, USA.

Sava Mintchev, School of Computer Science, University of Westminster, Harrow Campus, Horthwick Park, Harrow HA1 3TP, UK, s.m.mintchev@westminster.ac.uk.

A. Nikami, Dept. of Mathematical Engineering, University of Tokyo, 7-3-1 Hongo, Bunkyo-ku, 113 Tokyo, Japan, nikami@ipl.t.u-tokyo.ac.jp.

Steven D. Nolen, Nuclear Engineering Dept, Texas A&M University, College Station, TX 77843-3133, USA.

H. Ogawa, Dept. of Mathematical Engineering, University of Tokyo, 7-3-1 Hongo, Bunkyo-ku, 113 Tokyo, Japan, ogawa@ipl.t.u-tokyo.ac.jp.

Takashi Ohta, Japan Atomic Energy Research Institute, Nakameguro, Meguro-ku, Tokyo, Japan, takashi@koma.jaeri.go.jp.

John L. Pfaltz, Department of Computer Science, University of Virginia, Charlottesville, VA 22903, USA, pfaltz@cs.virginia.edu.

Dan Quinlan, Scientific Computing Group CIC-19, Los Alamos National Laboratory, Los Alamos, NM 87545, USA, dquinlan@lanl.gov.

Jürgen W. Quittek, ICSI, 1947 Center St., Suite 600, Berkeley, CA 94704, USA, quittek@icsi.berkeley.edu.

Ravi Ramamoorthi, Department of Computer Science, MS 256-80, California Institute of Technology, Pasadena, CA 91125, USA, ravir@cs.caltech.edu.

John V. W. Reynders, Advanced Computing Lab, Los Alamos National Laboratory, Los Alamos, NM 87545, USA, reynders@acl.lanl.gov.

Adam Rifkin, Department of Computer Science, MS 256-80, California Institute of Technology, Pasadena, CA 91125, USA, adam@cs.caltech.edu.

Bert Robben, K. U. Leuven, Department of Computer Science, Celestijnenlaan 200A, B-3001 Leuven, Belgium.

Yves Roudier, Electrotechnical Laboratory, Japan, iroudier@etl.go.jp.

Bill Saphir, NERSC, MS 50B-2239, 1 Cyclotron Road, Berkeley, CA 94720, USA, WCSaphir@lbl.gov.

Mitsuhisa Sato, Real World Computing Partnership, Tsukuba Mitsui Building 16F, 1-6-1 Takezono, Tsukuba-shi, Ibaraki 305, Japan, msato@trc.rwcp.or.jp.

Jonathan Schaeffer, Computer Science Department, University of Alberta, Edmonton, Alberta, Canada, T6G 2H1, jonathan@cs.ualberta.ca.

Tatiana Shpeisman, Xerox PARC, 3333 Coyote Hill Road, Palo Alto, CA 94304, USA.

Jeff Squyres, Laboratory for Scientific Computing, Department of Computer Science and Engineering, University of Notre Dame, Notre Dame, IN 46556, USA, squyres@cse.nd.edu.

Vaidy S. Sunderam, Mathematics and Computer Science Department, Emory University, 1784 N. Decatur Road, Atlanta, GA 30322, USA, vss@mathcs.emory.edu.

Duane Szafron, Computer Science Department, University of Alberta, Edmonton, Alberta, Canada, T6G 2H1, duane@cs.ualberta.ca.

Toshiyuki Takahashi, Information Science Department, University of Tokyo, 7-3-1 Hongo, Bunkyo-ku, Tokyo 113, Japan, tosiyuki@is.s.u-tokyo.ac.jp.

Yong-Tai Tan, DSO National Laboratories, 20 Science Park Drive, Singapore 118230.

Aslak Tveito, Department of Informatics, University of Oslo, P.O. Box 1080, Blindern, N-0316 Oslo, Norway, aslak@ifi.uio.no.

Henk Van Wulpen, K. U. Leuven, Department of Computer Science, Celestijnenlaan 200A, B-3001 Leuven, Belgium.

Todd L. Veldhuisen, Department of Systems Design Engineering, University of Waterloo, Waterloo, Ontario, Canada N2L 3G1, tveldhui@monet.uwaterloo.ca.

Pierre Verbaeten, K. U. Leuven, Department of Computer Science, Celestijnenlaan 200A, B-3001 Leuven, Belgium.

Matthias Weidmann, Institut für Informatik, Technische Universität München, D-80290 München, Germany, weidmann@informatik.tu-muenchen.de.

Boris Weissman, ICSI, 1947 Center St., Suite 600, Berkeley, CA 94704, USA.

Matthias Wilhelmi, RWCP Massively Parallel Systems GMD Laboratory Berlin, Rudower Chaussee 5, D-12489 Berlin, Germany, wilhelmi@first.gmd.de.

Akinori Yonezawa, Information Science Department, University of Tokyo, 7-3-1 Hongo, Bunkyo-ku, Tokyo 113, Japan, yonezawa@is.s.u-tokyo.ac.jp.

Xingbin Zhang, Departmentof Computer Science, University of Illinois at Urbana-Champaign, Urbana, IL 61801, USA, concert@red-herring.cs.uiuc.edu.

Author Index

Springer
and the
environment

At Springer we firmly believe that an international science publisher has a special obligation to the environment, and our corporate policies consistently reflect this conviction.

We also expect our business partners – paper mills, printers, packaging manufacturers, etc. – to commit themselves to using materials and production processes that do not harm the environment. The paper in this book is made from low- or no-chlorine pulp and is acid free, in conformance with international standards for paper permanency.

 Springer

Lecture Notes in Computer Science

For information about Vols. 1–1269

please contact your bookseller or Springer-Verlag